ET, ARE YOU OUT THERE?

The question of intelligent life on other worlds

Surendra Verma

This book is a completely revised, updated and expanded edition of the author's *Why Aren't They Here? The Question of Life on Other Worlds,* published in hardback in 2007 and paperback in 2008 by Icon Books, UK

This edition first published in 2022

Cover photo: Courtesy of NASA

ISBN: 9798828860104

CONTENTS

ALIENS, SHOW YOURSELF, PLEASE

Well, it suddenly appears in the sky, dazzles lonely little earthlings and then disappears. The faithful gaze in silent wonderment and yearn for a glimpse of the little green men – oops, little green persons – at the helm of the UFO. Sceptics scoff at this naïve delusion and call them out for their silliness. To them, UFOs are not worthy of comparison with the scientists' search for extraterrestrial intelligence (SETI). One day we may eavesdrop on an interplanetary ET conversation is no longer a crazy idea as aliens visiting us. We cannot be alone in the vast universe.

Aliens and UFOs are not the same things, but both fire up our imagination. Everyone loves ET, and we want to believe friendly, not frightening, aliens exist. One day we may receive a message from an extraterrestrial intelligent life from a distant world. Will we ever meet an alien stepping out of a UFO? Highly unlikely.

A single UFO of extraterrestrial origin will inspire fear among humans that the aliens could be malevolent, but they could also be benevolent, says American astrophysicist Avi Loeb. 'Here's hoping that if and when the time comes,' he says, 'we will learn that their intent is wise and well-meaning.' Pray the distinguished professor is right.

A long-awaited US intelligence document released in 2021 has failed to assure that the aliens are not watching us. To the delight of UFO fans, aliens watching us have fired up their flying saucers and soon be visible only to believers. As the report did little to endorse or dismiss UFO theories, conspiracy pundits are pepped up to spin out more conspiracy theories damning the government once again for lying to us and hiding the hard evidence of visitors from other worlds. SETI enthusiasts can now double down on deciphering messages from intelligent aliens as the report has not ruled out aliens.

One cryptic message did arrive on 30 June 1908 from ETs inhabiting a giant planet orbiting the star 61 Cygni, about 11 light-years away. The violent volcanic eruption of Krakatoa in August 1883 generated strong radio waves. Cygnian scientists received this signal 11 years later. But they misread the signal as greetings from a distant civilisation. Following the age-old Cygnian custom, the courteous scientists decided to send a return message. As their laser technology was more advanced than radio technology, they directed a laser beam at our planet. Unfortunately, the well-meaning scientists made another mistake. They misjudged Earth's distance and fired a powerful laser beam that zapped the Tunguska taiga in remote Siberia, wiping out millions of trees. This 'extra strong' Cygnian message was all Greek to the local Evenk people; they did not have the required technology to read their greeting card from the stars. This is one of many explanations for the so-called Tunguska event. More on visitors and whispers from space later; first, the US report.

UFOs are so infrequent that they are unique to most observers. In that sense, they are truly encounters of the unidentified kind, but they are not encounters with extraterrestrials. 'U' in UFO simply means 'unidentified'; it does not suggest 'something beyond this world'. To remove the stigma, the term UFO carries, the report from the US Department of Defense Unidentified Aerial Phenomena Task Force calls them UAP or unidentified aerial phenomena. This name change may encourage pilots to report new sightings and scientists to study them.

The report reviewed 144 sightings since 2004 of unknown objects moving without observable propulsion or rapid acceleration unknown to the US and beyond the capabilities of other nations. 'Some UAP appeared to remain stationary in winds aloft, move against the wind, manoeuvre abruptly, or move at considerable speed, without discernible means of propulsion,' the report says. The report concludes that there is no evidence that any UAPs involve secret American weapons programs, unknown technology from China or Russia or extraterrestrial visitations. However, the report admits that there are objects in the sky that cannot be explained.

The inconclusive UAP report – to dedicated believers, they will always be UFOs – has encouraged me to look at again the question of intelligent life on other worlds. Our contradictory views on aliens and their flying machines continue, but UFOs are unlimited fun to both believers and sceptics.

UK's *Daily Mail* newspaper described my 2007 book, *Why Aren't They Here? The Question of Life on Other Worlds* (Icon Books, UK) as 'a calm, intelligent and witty survey of the history of humankind's search for extraterrestrial life.' I hope this new attempt lives up to the same standard and, as before, is 'in-your-face, direct and beautifully written' (as *BBC Focus* magazine described it). As before, my approach is sceptical and light-hearted yet rigorously scientific. However, reading this book requires no background in science, only a curious mind. Like *Alice in Wonderland*, the trip down the alien rabbit hole will become 'curiouser and curiouser' as you continue reading.

CHAPTER 1

EARTHLING FOLLIES

There is a flying saucer in your backyard

The first UFO in the shape of a flying saucer entered our world when, on 8 July 1947, the *Roswell Daily Record* broke the news of a cosmic encounter in New Mexico. The story, headlined, 'RAAF captures flying saucer on ranch in Roswell region,' was based on a press release issued by Roswell Army Airfield (RAAF). When sheep rancher Mac Brazel was making rounds at a ranch 137 kilometres (85 miles) west of Roswell, he found some wreckage consisting mainly of rubber strips, wood sticks, tinfoil, plastic, tape with some strange markings that resembled 'hieroglyphics', and very tough paper. The unusual nature of the debris struck Brazel. After a few days, he drove into Roswell, where he reported the incident to the Sheriff, who reported it to Major Jesse Marcel, Intelligence Officer at RAAF.

The Army closed off the debris site while the wreckage was being cleared. The officers thought that they had found a flying saucer. They shipped the debris to Air Force General Roger Ramey for examination. But in the meantime, Colonel William Blanchard, the Commander at Roswell, issued a press release saying that the wreckage of a flying saucer had been found. The news caused a sensation around the world, but it was short-lived. Within hours General Ramey called in the local press and announced that RAAF had mistakenly identified the remains of a weather balloon as the wreckage of a flying saucer. The next day, the *Roswell Daily Record*'s banner headline proclaimed: Gen. Ramey Empties Roswell Saucer.

Another story that the paper did not publish, but some town folks knew about came from witnesses who had seen the wreckage. They described the wreckage as three flying saucers, circular in shape with raised centres, about 15 metres (50 feet) in diameter. Each one was occupied by three alien bodies of human form but only a metre (three feet) tall, with bluish skin colouration and no ears, hair, eyebrows or eyelashes. The Air Force explained these 'aliens' as dummies dropped from high-altitude balloons to study the results of the impact. And that was the end of the excitement.

No one ever talked about this episode, at least until the publication in 1980 of a book, *The Roswell Incident*, which came to the dramatic conclusion that there had been a cover-up of cosmic proportions. In 1988, another book, *UFO Crash at Roswell*, claimed that the US government had found and removed the remnants of the UFO crew — several little alien bodies. These two books were the genesis of a UFO myth and a conspiracy theory — that the government had conspired to cover up the fact that an alien ship had landed at Roswell. The truth is far less exotic: what happened was that people who saw the dummies mistook them for aliens.

When flying saucers became UFOs and wowed the world

The term 'UFOs' (unidentified flying objects) was coined in 1955 by Edward Rupplet, a US Air Force officer who worked on one of the first attempts to understand strange objects' sightings in the sky. He considered the term 'flying saucer' vague since many sightings had very natural explanations while others did not. He wrote in a report, 'People want to know the facts. But more often than not, those facts have been obscured by secrecy and confusion, a situation that has led to wild speculation on one end of the scale and almost dangerously blasé attitude on the other.' The secrecy and confusion still cloud any government effort to explain UFOs.

Most UFO fans believe that extraterrestrial intelligent beings are visiting Earth. They also think governments are covering up this fact because they know it would trigger a panic; governments are afraid of admitting something beyond their control.

Most of these UFO fans have also claimed to have seen a UFO. What have they seen? Planes, jets, helicopters, balloons, strange flocks of birds. Unusual light patterns are caused by astronomical and meteorological phenomena. Optical illusions are caused by smoke and dust. Psychological delusions. Deliberate hoaxes.

Another explanation is a mass suggestion. The power of a placebo is well known. A sugar pill can fool the mind into thinking the problem is being taken care of. Intriguingly, you can experience a placebo effect even if you do not believe in it. The placebo effect also works in reverse. In 1947, *The New York Times* published a story saying that 'flying disk fail to stir the Air Forces' in response to a businessman's claim that he saw nine mysterious objects – as big as aeroplanes – racing over Washington State's Cascade Range. Sever other persons in widely scattered localities *later* said they had glimpsed similar objects. The *day after the story* appeared in the paper, flying saucers were spotted over New York. The 'flying saucer' though still adept at eluding the most powerful telescopes in America continued to flash in increasing numbers and variety before the goggling eyes of rooftop and roadside amateurs, the newspaper said. Incidentally, it was the first mention of the term 'flying saucer' in the venerable newspaper.

'Disks' Soar Over New York, Now Seen Aloft in All Colors

By MURRAY SCHUMACH

The "flying saucer," though still adept at eluding the most powerful telescopes in America, continued yesterday to flash in increasing numbers and variety before the goggling eyes of rooftop and roadside amateurs.

New York and other Eastern states, hitherto oblivious to the strange bodies in the sky, suddenly found they were not immune, according to the latest reports to harassed policemen and astronomers. The Associated Press said that thirty-nine states, plus the District of Columbia and a part of Canada were playing host to the heavenly disks.

Despite the humorous skepticism of scientists and military experts, the latest flock of rumors showed increasing imagination. No longer, for example, were the disks just white. In some cases they were in technicolor, with orange the predominant hue.

These explanations do not convince many people who have had vivid memories of close encounters with UFOs. Scientists say that some UFO events are attributable to physical, electrical and magnetic phenomena in the atmosphere. These events create regions of electrically charged plasma which appear as bright, fast-moving objects to observers. They are probably caused by a meteorite entering the atmosphere, neither burning up completely nor impacting, but forming buoyant plasma. Sometimes the field between particular charged buoyant objects forms an area, often triangular, which does not reflect light. Some UFOs are described as black spaceships, usually triangular, and up to hundreds of metres in length. The events cannot be detected by radar.

The proximity of plasma related fields can adversely affect a vehicle or person. It has been medically proven that local electromagnetic fields can cause responses in the temporal lobes of the human brain. They result in the observer sustaining (and later describing and retaining) their own vivid but incorrect description of what is experienced.

The celebrated American astronomer Carl Sagan of *Cosmos* (the renowned book and TV series) fame gives an interesting explanation: 'If each of a million advanced technical civilisations in our galaxy launched at random an interstellar spacecraft each year, our solar system would, on average, be visited once every 100,000 years.'

Sagan, who died in 1996, employs his Santa Claus hypothesis to explain UFOs. The idea supports that in eight hours or so on 24–25 December of each year, an elf visits 100 million homes in the United States. If this elf spends one second per house, he must spend three years just filling the stockings in all the places. Even with relativistic reindeer, the time spent in 100 million homes is three years and not eight hours. 'We would then suggest that the hypothesis is untenable,' he says. 'We can make a similar examination, but with greater certainty, of the extraterrestrial hypothesis that holds that a wide range of UFOs viewed on Earth are space vehicles from planets of other stars.'

The UFO genie is now out of the bottle. To put it back in the bottle is not a task for us earthlings. Only aliens can do it when they pop out of a UFO and make it an IFO, an identified flying object.

Abducted by aliens!

There is no hard evidence that there is intelligent life beyond Earth, but still, some people believe that aliens have visited the planet. Some even believe that aliens abducted them. The abduction stories of these people have a remarkable degree of similarity and consistency: Your naked body is floating on a high-tech table inside a round bright saucer-like object and being subjected to a painful, invasive medical examination by aliens with large heads, slanted wrap-around black eyes, either grey, white or green skin and no hair or nose.

If aliens ever abduct you:

Encounter: You will be either in your bed or car, usually at night. If you are in your bed, you suddenly wake up and see non-human figures coming through the walls of your room or standing near your bed. You may even see a spaceship outside your window. If you are in your car, you see the car being pulled to the side by a bright object. You see strange beings exiting from a spaceship and coming towards you.

You are unable to move or speak. You see flashing lights and hear buzzing sounds.

The aliens are usually about 120 centimetres (4 feet) tall. Their bodies are thin and spindly with huge heads, slanted wrap-around black eyes, grey skin and no hair or nose. They don't talk but communicate with you telepathically with their large eyes.

Abduction: You are taken to the spaceship. You don't walk; you float or are carried up by a light beam. The inside of the spaceship looks like a high-tech doctor's examination room full of tables on which other humans lie.

Examination: You are subjected to a painful medical examination. The aliens probe you by inserting instruments into virtually every body part. Sometimes they place tiny implants in your body, especially the nose. No person has ever been found with an actual implant. Most abductees claim that they lost the nose implant when they sneezed.

Tour: The aliens may give you a tour of their spaceship or show some alien artefacts. No person has ever brought back any extraterrestrial souvenirs from their flying saucer tour.

Return: It's all over within a brief time. You are back in your bed or car. You are confused and scared. You may find puzzling 'surgical' scars or cuts on your body, and you don't remember how you got them.

UFOs: Many stories have associations with UFOs. The abductees may not have seen the UFO, but they usually read or see on the television the next day that a UFO passed over where they were when they had the experience.

Location: Most incidents happen in the United States. No one knows why the aliens always pick up Americans.

The alien-abduction phenomenon began with the case of Betty and Barney Hill. Late one night in 1961, when Betty and her husband Barney returned to their New Hampshire home from a holiday in Canada, they noticed a bright light following their car. The light became brighter and brighter until it was visible as a spaceship. Their car stalled, and the spaceship landed nearby. The aliens came out and took them inside the spaceship, where they were subjected to a medical examination. Betty was given a brief tour of the spaceship and when she asked where the aliens were from, she was shown a star map.

When they 'awoke', the couple continued their drive home without any memories of the incident. Weeks afterwards, they started experiencing nightmares and went to see a psychiatrist. Under hypnosis, they 'remembered' having been abducted by aliens and subjected to painful probing of their bodies; Betty also drew the star map shown by the aliens. A few years later, a UFO researcher claimed that Betty's vague star map resembled Zeta Reticuli, a binary star system located in the constellation Reticulum, 39 light-years away.

Betty and Barney's story became highly popular when John Fuller told it in his

1966 book, *The Interrupted Journey*. It was later made into the television movie *The UFO Incident*. The book and the movie started a new genre, and the publishing and entertainment industries continue to milk this cash cow.

At least one eminent scientist took these strange happenings seriously. John Mack, a Pulitzer Prize-winning psychiatrist at Harvard University who died in 2004, was ridiculed by the scientific community when he claimed in his 1994 book, *Abduction: Human Encounters with Aliens*, that there was no psychological explanation for the phenomenon. He argued that abductees – he worked with 200 of them and called them 'experiencers' – were not crazy, and their experiences were genuine:

'It's both literally, physically happening to a degree, and it's also some kind of psychological, spiritual experience occurring and originating perhaps in another dimension … I would never say, yes, there are aliens taking people. But I would say there is a compelling, powerful phenomenon here that I can't account for in any other way; that's mysterious. Yet I can't know what it is, but it seems that it invites a deeper, further inquiry.'

Not you; it's your brain that aliens have abducted

People who believe in alien abduction tend to fantasise and hold to unusual beliefs and ideas. They also believe in things like ESP, astrology, tarot, channelling, auras and crystal therapy. Susan Clancy, author of *Abducted: How People Come to Believe Aliens Kidnaped Them*, says these people are not crazy. Still, they have in common a rash of disturbing experiences for which they are seeking an explanation. For them, alien abduction is the best fit. 'Many of us long for contact with the divine, and aliens are a way of coming to terms with the conflict between science and religion,' she says. 'I agree with Jung: extraterrestrials are technological angels.'

Some psychologists have tied the phenomenon to sleep paralysis, a condition where the usual separation between sleep and wakefulness gets out of synchronisation. Sleep occurs when the body is in the dream phase of sleep, and it disconnects from the brain. The brain is awake or semi-awake, but the body cannot move. At that point, the sleeper often 'sees' shadowy creatures, 'experiences' levitation and 'feels' painful sensations throughout the body, explains Kazuhiko Fukuda, a Japanese psychologist. Such experiences match with the accounts of people claiming to be victims of alien abduction.

There are other psychological explanations as well. Some psychologists believe that alien abductions and different mystical and psychic experiences may be linked to excessive bursts of electrical activity in the temporal lobes. These lobes – one on each side of the brain, located near the ears – control hearing, speech and memory. Some argue that alien abduction may be disguised memories of sexual abuse. Or, they may be false memories. People can and do make powerful memories, and these memories can take on a life of their own. All abduction stories have been recalled under hypnosis. Hypnosis makes people susceptible to creating memories of things that were suggested to them or things they just imagined.

US psychologist David V. Forrest offers two medical hypotheses. His 'strong' idea is that the abductees are recovering memories of actual surgery – the memories may be a factual recall of the operating room before losing consciousness, or they could be memories from childhood filtered through childhood amnesia. His 'weak' hypothesis is that abductees are confabulating media conceptions of aliens with images of surgical and medical procedures generally, images that they may or may not have experienced personally.

Whatever may be the correct explanation for alien abduction accounts, why are their stories so similar? Anthropologists point out that individual illusionary experiences conform to cultural patterns. Alien abductions occur mainly in the United States, where people are familiar with alien references through supermarket tabloids, books, movies and TV shows. Abduction accounts appeared in 1962 when alien abduction also began appearing on TV and in movies. For example, during the witchcraft craze in medieval Europe, many people reported being carried away by witches on broomsticks and being seduced by demons. Today's counterparts talk of being picked up by flying saucers and forced to perform various forced sex acts by their bug-eyed kidnappers.

The alien-abduction stories may seem fantasies to us, but they play a role in the psychological lives of those who believe in them.

Chariots of the gullible, not the gods

Scientists are still arguing about the existence of microbial life on other worlds, not to speak of intelligent life. Still, in 1968 Erich von Daniken, a mild-mannered Swiss hotel manager with no scientific training, published a book in which he claimed that intelligent aliens visited our planet in prehistoric times. Published in English in 1969 under the title, *Chariots of the Gods?,* the book became a bestseller and inspired a whole 'ancient astronauts' industry of countless books, documentaries and films.

The fad has almost run its course, but not completely. The book is still in print, and tens of thousands of websites are now devoted to its startling but poorly researched ideas.

With the zeal of a UFO convert, von Daniken professed that we are the descendants of astronauts from outer space who came to Earth about 10,000 years ago; our gods are simply the astronauts that visited us in the past. He said mythical stories of many lands filled with godlike visitors from the sky arriving in fiery chariots support this assertion.

He argued that 'unsolved mysteries of the past', such as the moving of large stones for building pyramids and the great carved heads of Easter Island, could be explained by knowledge learned by our ancestors from visiting aliens. He suggested that ancient patterns of triangles, rectangles and trapezoids, huge spiders, monkeys, birds, fish and reptiles laid out with equal precision across 50 kilometres (31 miles) of Peru's Nazca plain were landing strips for 'ancient astronauts'. Some Nazca patterns are indeed so large that they could only be recognised from the air, but there is a down-to-earth explanation for them: they were probably drawn by Nazca's astronomer-priests to mark the passage of seasons.

Archaeologists have questioned von Daniken's research techniques and reasoning. They say that, of course, there are things which are still mysteries from the past. Still, von Daniken's freewheeling speculations based on bits and pieces of archaeological 'facts' have failed to present a cohesive and convincing argument.

Planets colliding on the edge of science

In about 1500 BC, a comet was ejected from Jupiter. It came close to Earth and Mars before settling into an orbit between Earth and Mercury and becoming the planet Venus. As Earth passed through the comet's tail, it caused a range of events: meteoric bombardment started widespread fires; falling meteoric dust turned rivers blood red and plunged the world into the darkness that lasted four days; Middle East oil fields were created as large quantities of petroleum fell from the sky, and edible carbohydrates (manna) fell on Earth. For hundreds of years, multiple collisions occurred between Earth, Venus and Mars, which ended in 700 BC. Earth suffered extreme geological changes from these close passes of Venus and Mars, causing catastrophic events such as volcanic eruptions, earthquakes, global floods and the formation of new mountain ranges. The evidence for these cosmic collisions can be found in myths and legends of all cultures; for example, in Biblical stories such as Noah's flood, the parting of the Red Sea, Joshua making the Sun standstill, and people eating manna from the sky.

All these claims of violent catastrophes and many more appeared in a book, *Worlds in Collision*, published in 1950. Its author, Immanuel Velikovsky, was a Russian-American psychiatrist who died in 1979. The book unleashed a torrent of arcane information. Apparently, it impressed book critics. They gave it highly favourable reviews in popular magazines and newspapers, which made the book hugely popular (it's still in print). Astronomers were not so impressed with this cataclysmic concept of world history.

All Velikovsky's theories have been thoroughly rejected, but there are still many dedicated followers. Writes David Morrison, an American space scientist, in 'Velikovsky at Fifty', an article published in 2001 in the *Sceptic* magazine: 'There is ample evidence that Velikovsky was little more than a crank, something that was evident to the astronomers from even a cursory look at his book.'

Reflected Brian Toon, an American atmospheric scientist: 'Velikovsky influenced me by showing how the public is so easily fooled by pseudoscience.'

Moon hoax: Americans didn't land on the Moon; batmen did

You have probably heard about the conspiracy theory that the Apollo 11 moon landing was an elaborate hoax. Conspiracy theorists' websites and blogs revved up in 2001 when Fox TV aired a program 'Conspiracy Theory: Did We Land on the Moon?', arguing that NASA technology in the 1960s wasn't up to the task of actual Moon landing; instead, it faked the Moon landing in movie studios. A witty comment on the NASA website should silence the pedlars of Moon hoax stories who ignore the overwhelming evidence: 'Fortunately, the Soviets didn't think of the gag first. They could have filmed their fake Moon landing and really embarrassed the free world.'

In the early 19th century, when there were no conspiracy theorists around, but hoaxes abounded, a clever Moon hoax claimed that a famous British astronomer had discovered intelligent life on the Moon. Here's the story of that hoax.

On 25 August 1835, a New York newspaper, *The Sun*, ran a front-page story with the headline 'great astronomical discoveries lately made by Sir John Herschel'. The report, purporting to be a reprint from a supplement of the *Edinburgh Journal of Science*, described fantastic sights of the Moon viewed by Herschel from his new telescope at the Cape of Good Hope. The story, which continued in instalments over a few days, described the lunar landscape of vast forests, inland seas and lilac-hued 'very slender pyramids, standing in irregular groups, each composed of about 30 or 40 spires'.

As *The Sun*'s sales soared from 8,000 to 19,360 copies, its readers were introduced to *Vespertilio-homo* or 'bat-man' on the Moon: 'Certainly they were like human beings ... They averaged four feet in height, with short and glossy copper coloured hair and had wings composed of a thin membrane ...'

At the time of the publication of the articles, John Herschel, one of the most famous scientists of his time, was in Cape Town surveying the southern skies (John was the son of William, who discovered Uranus in 1781, the first recorded discovery of a planet in human history). His only known comment is in a letter to his Aunt Caroline: 'I have been pestered from all quarters with that ridiculous hoax about the Moon – in English, French and German.'

The author of the articles, the British-born journalist Richard Adams Locke, later claimed that the story was a satirical piece written to show the gullibility of Americans on the question of extraterrestrial life.

Crops circles: reaping by aliens, humans or nature?

The term 'crop circles' refers to enormous geometrical patterns that appear somewhat mysteriously, usually overnight, in fields of wheat, barley, rye, oats and other cereal crops. The plants are rarely cut or damaged, but their stalks are swirled and flattened.

Crop circles have been reported since the 1960s. Still, the phenomenon attracted worldwide widespread media attention in the northern summer of 1980 when several circles, each about 18 metres (60 feet) in diameter, appeared randomly in a field of oats in the rolling Wiltshire countryside in southern England. Since then, thousands of circles have been reported, some in other countries but mainly in southern England. As their numbers have increased, so have the complexity and size of their designs. Relatively simple circles of the 1960s have evolved into elaborate pictograms and shapes (sunflower, scorpion, spider web, the Buddhist wheel of Dharma, the Pythagorean symbol of wellbeing) and complex mathematical and fractal patterns (ten digits of pi, Euclid's theorems, Ptolemy's theorem of chords, a vortex of logarithmic curves). A formation that appeared in 2001 (in Wiltshire, of course) covered about five hectares (12 acres) and was more than 243 metres (800 feet) across.

Most of the recent crop circles are the work of hoaxers prompted by media attention. But those who study crop circles (they call themselves 'cerealogists') reject this mundane idea and offer other exotic explanations: (1) they are the work of intelligent aliens as most crop circles appear close to UFO sightings; or (2) they are caused by spiritual energy or nature spirits. There is a conspiracy theory: they are formed by testing a secret microwave weapon the government has developed and do not want to tell the public about it.

Scientists, however, offer simple explanations. Precisely a century before the appearance of the famous Wilshire circles, an English amateur scientist J. Rand Capron described strange circular spots in a wheat field in a letter to the journal *Nature* on 29 July 1880, 'Examined more closely, these all presented much the same character, viz., a few standing stalks as a centre, some prostrate stalks with their heads arranged pretty evenly in a direction forming a circle round the centre, and outside these, a circular wall of stalks which had not suffered … and I could not trace locally any circumstances accounting for the peculiar forms of the patches in the field, nor indicating whether it was wind or rain, or both combined.' He concluded that the phenomenon was probably caused by 'some cyclonic wind action'.

Modern scientists also suggest that they are formed by an atmospheric vortex, some kind of swirling air current like a tornado. Others postulate that this vortex must be electrified to make such complex patterns. Critics say that even an electrified vortex cannot cut such precise patterns on the ground.

If all crop circles are not fabricated, scientists have a challenging task. The problem is that most 'credible' scientists do not like to involve themselves in weird and off-beat things.

The following 1687 woodcut pamphlet entitled 'Mowing-Devil' tells a tale about an English farmer who asked the village mower to cut down his crop of oats. When the mower demanded too much money, the farmer replied that the devil should mow it rather than you. The whole field was cut down mysteriously the same night. The illustration shows the devil with a scythe mowing in a circular design, not just a small section as in the case of modern crop circles.

The Mowing - Devil:

Or, Strange NEWS out of Hartford - shire.

Being a True Relation of a Farmer, who Bargaining
with a Poor *Mower*, about the Cutting down Three Half
Acres of *Oats*, upon the *Mower's* asking too much, the *Farmer* swore, That the *Devil* should Mow it, rather than He.
And so it fell out, that that very Night, the Crop of *Oat*
shew'd as if it had been all of a Flame, but next Morning
appear'd so neatly Mow'd by the *Devil*, or some Infernal Spirit, that no Mortal Man was able to do the like.
Also, How the said *Oats* ly now in the Field, and the Owner
has not Power to fetch them away.

Licensed, August 22th, 1678.

Thirsty aliens

About 7.14 a.m., 30 June 1908. The Central Siberian Plateau near the Stony
Tunguska River, a remote and empty wilderness of swamps, bogs and hilly pine
and cedar forests. Not a soul in sight for scores of kilometres. The eerie silence is
punctuated by the shuffle of the hoofs of reindeer grazing in the morning sun.

Suddenly a blindingly bright pillar of fire, the size of a tall office building, races across the clear blue sky. The dazzling fireball moves within a few seconds from the south-southeast to the north-northwest, leaving a thick trail of light hundreds of kilometres long. It descends slowly for a few minutes and then explodes above the ground. The explosion lasts only a few seconds, but it is so powerful that it can be compared only with an atomic bomb – 1,000 Hiroshima atomic bombs. A dark mushroom cloud almost reaches space.

The explosion flattens an area of remote Tunguska forest bigger than Greater London, stripping millions of ancient trees of leaves and branches, leaving them bare like telegraph poles and scattering them like matchsticks. A black rain of debris and dirt follows.

If it were not a giant meteorite or meteorite crater, then what caused the great Siberian explosion? The controversy about the Tunguska fireball continues to this day, and there is no shortage of attempts to explain the cataclysmic explosion.

A bizarre idea is that the explosion was caused by a 'cosmic visitor' – an extraterrestrial spacecraft, cylindrical in shape and propelled by nuclear fuel. Because of a malfunction, the spacecraft hit Earth, and within a fraction of a second, the spacecraft and its occupants were vaporised in a blinding flash of light. The ETs had come to collect water from Lake Baikal, which is the largest volume of surface freshwater. Apparently, the ETs were from a parched planet and very, very thirsty.

UFO fans prefer the theory that a faulty flying saucer was deliberately crashed into the sparsely populated Siberia by its considerate extraterrestrial crew to save human lives. Some add another twist: the explosion was caused when a spaceship's engine blew up. The spaceship had left two thousand years earlier and was returning home but missed the runway.

Hollow Earth: aliens in the deep

'EARTH HAS A hollow interior' was first proposed as a scientific theory by the English astronomer Edmund Halley (of comet fame) in 1691 at a meeting of the prestigious Royal Society. Halley, who had helped in the publication of Newton's *Principia* in 1687, argued that Earth was composed of four shells: an outer shell 800 kilometres (500 miles) thick, then two inner shells of diameters comparable to Mars and Venus and the solid inner shell about the size of Mercury. He also suggested that the atmosphere between the two inner shells was illuminated, and therefore the shells were capable of bearing life. When a brilliant *Aurora borealis* (beautiful display of coloured lights in far northern skies) was seen in 1716, he thought it was caused by the escape of glowing gas in the inner shells through a hole in the North Pole. Obviously, *Aurora australis* (display in far southern skies) was unknown to him as Australia had not yet been discovered.

In the 18th century, the famous Swiss mathematician Leonhard Euler expanded Halley's idea, suggesting that Earth was completely hollow, could be reached through holes in the North and South Poles, and had its own sun 960 kilometres (600 miles) wide.

The idea became highly popular when in 1913, Marshall B. Gardner, an American engineer, published a little book, *Journey to the Earth's Interior. He* claimed that the extinct mammoth discovered in Siberia in 1846 came from the interior of Earth. Eskimos also originated from the deep, as shown by their legends of a warm land with perpetual daylight. In 1914 the United States Patent Office accepted Gardner's application for a patent on his hollow-earth theory.

In the 1930s, the Nazis, who were suckers for silly ideas, also picked up the concept of hollow Earth. There is no proof, but it has been alleged that expeditions were sent to Antarctica and Tibet to meet the aliens in the deep. And there is a bizarre suggestion that Hitler and some other Nazis escaped into the hollow Earth. True or not, their remains are now definitely there.

The geological truth, however, is not as exciting. Geologists tell us that Earth is made up of three main layers: (1) the outer layer, the crust; (2) below the crust is the mantle, a thick shell of molten rock separating the crust from the inner core; and (3) the core is extremely hot and dense and is about 2,900 kilometres (1,800 miles) below the surface.

You will not find any Nazis or other alien creatures in Earth's centre unless you're reading Jules Verne's 1864 classic novel, *Journey to the Centre of the Earth*.

Why are we suckers for weird beliefs?

Alien abduction, ancient astronauts, astrology, crystal healing, ESP, magnetic therapy, quantum healing, UFOs … the list of absurd things is long, and the number of people who not only believe in such rubbish but vigorously defend them is exceptionally large. There is absolutely no evidence, scientific or otherwise, to support their beliefs. Ockham's razor – the simplest explanation is most likely to be right – is alien to believers in UFOs; science is crystal clear to practitioners of crystal healing.

Another overwhelming weird idea is quantum healing. Quantum mechanics is mind boggling even to scientists. Super-brilliant physicist Richard Feynman once admitted that it 'appears peculiar and mysterious to everyone – both the novice and the experienced physicists.' What does a theory of matter have to do with mind and body? This anomaly has not stopped New Age gurus to draft 'quantum mechanics' in the service of their so-called holistic healing practices. Quantum healing is a mishmash of ideas collected from quantum mechanics, ancient Hindu metaphysics and Ayurveda, the ancient Hindu system of medicine. This quantum quackery lacks robust – even weak – evidence, yet it appeals to millions of intelligent, educated people. Using the word 'quantum' in the context of healing would have made the German physicist Max Planck cringe.

Though Planck's theory proposed in 1900 had predicted the existence of quanta or photons – a photon is an elementary particle representing a quantum of light or other electromagnetic radiation – he was doubtful about their reality. He once visited a physics lab where he saw an apparatus that counted photons by audible clicks in action. He stood silently for a while and just listened. Then he smiled and murmured, 'So they do exist.' Does quantum healing exist? We must apply the same scepticism to quantum healing as Planck applied to his quantum theory.

Why does the human brain allow and even encourage beliefs that defy reason?

The answer probably lies in your parietal lobe – a mass of tissues at the top of the brain which processes sensory input and distinguishes where the body ends and the material world begins. During intense prayer or meditation, the parietal lobe powers down. Unable to find the dividing line between self and the world, you experience the sense of having lost your worldly moorings. You feel connected to the 'other world'. The parietal lobe's ability to go quiet may encourage other beliefs that bring a sense of connection.

Brain imaging these days can reveal many mental acts. A trio of American neuroscientists – Sam Harris, Sameer Sheth and Mark Cohen – just did that to find out how the brain differentiates between belief and disbelief. Their ingenious and straightforward experiment involved presenting a series of written statements to participants while they were in the fMRI scanner. The statements were designed to be clearly true (belief), false (disbelief) or doubtful and were from seven categories: mathematical, geographical, semantic, factual, autobiographical, ethical and religious; for example:

- 62 can be evenly divided by 9
- Senegal borders Guinea
- 'Devious' means 'friendly'
- Most people have 10 fingers and 10 toes
- You have two sisters
- It is bad to take pleasure in another's suffering
- There is probably no actual Creator God

The results were fascinating. Response time to true statements was much shorter than response times to both false and doubtful statements; however, there was no difference in response times between false and doubtful statements. Response times to acceptance of belief were similar, whether the participants made a judgment in the highly emotional areas of ethics or religion or neutral area of mathematics.

Different brain regions frequently lit up when responding to belief and disbelief statements. Factual statements activated areas of the prefrontal cortex, which are thought to play a role in decision-making, memory and fear. In contrast, false statements showed increased activity in the anterior insula, which helps to process fear, disgust and reactions to bad smells. This finding suggests two distinct brain systems for belief and disbelief. The researchers suggest that belief or acceptance of a proposition as true has a pleasant and rewarding emotional tone. Disbelief or rejection of a proposition, on the other hand, is often associated with a feeling of discomfort and an urge to avoid 'untruth'. 'When someone says something, you disbelieve; it has a kind of emotional tone,' says Harris. 'Rejecting someone's statement as illogical or incompatible feels like something.'

Clay Routledge, a professor of psychology at North Dakota State University, offers a different but not conflicting explanation. Studies show that interest in supernatural and paranormal phenomena is driven by the same cognitive processes and motives that inspire religion. 'The less religious people are,' he says, 'the more likely they are to endorse empirically unsupported ideas about UFO, intelligent aliens monitoring the lives of humans and related conspiracies about a government cover-up of these phenomena.' From 2007 to 2014, the number of Americans who report being absolutely confident God exists has dropped from 71 per cent to 63 per cent. It's not surprising then that around two-thirds of Americans hold supernatural or paranormal beliefs of some kind.

Believing or doubting something controls our behaviour and emotions. While doubt tends to inhibit action, beliefs make it easier to arrive at a decision and act on it. Michael Shermer, the founding publisher of *Skeptic* magazine, states that once we form beliefs, we keep and reinforce them through five cognitive biases that distort our precepts to fit belief concepts:

- *Anchoring bias*: Relying too heavily on one piece of information.

- *Authority bias*: Valuing too much the opinions of authority.

- *Belief bias*: Accepting an argument on the believability of its conclusion.

- *Confirmation bias*: Seeking supporting evidence but ignoring discomforting evidence.

- *In-group bias*: Placing more value on the beliefs of our peers.

Should you worry if someone you know believes in weird things? 'It can cause no harm,' says Stephen Law, author of *Believing Bullshit: How Not to Get Sucked into an Intellectual Black Hole*. 'But the dangers are obvious when people join extreme cults or use alternative medicines to treat serious disease.'

Common sense and science: a complex relationship

'Common sense is the collection of prejudices acquired by age 18.' This famous quotation is attributed to Einstein, but its accuracy is questionable. What is common sense, anyway? It may be called ordinary, nonspecialised knowledge, which leads someone to make a sound judgment. American cultural anthropologist Lathel F. Duffield wants to add 'common' to this meaning and calls common-sense 'a judgment or opinion shared by members of a group'. He argues that this sharing makes common sense, a cultural pattern that is a part of the overall cultural configuration of society. Therefore, he says, common sense is a learned way of thinking.

This learned way could also lead to prejudices, which are undesirable because they result in unfair treatment of individuals. Common sense tells us that the Sun moves around Earth. In the 16th century, when the Polish astronomer Copernicus rejected this commonly held wisdom, he was ridiculed as a fool who wanted to turn the whole science of astronomy upside-down. Since then, common sense has evolved and now anyone who believes that Earth is the centre of the solar system would be ridiculed as a fool. When common sense fails to evolve, it becomes ignorance, prejudice, and even bigotry. We should not confuse common sense with prejudice.

The advent of quantum mechanics in the early 20th century introduced phenomena beyond our senses' capabilities (even Einstein was 'spooked' by quantum entanglement). Evolution's hardware and software 'wired' in humans had reached their peak. Science couldn't be simply guided by common sense. 'A classic example of the limitations of our neural wiring is the inability to picture more than three dimensions,' says American theoretical physicist Leonard Susskind. 'Physicists have had no choice but to rewire themselves. Where institution and common sense failed, they had to create new forms of intuition, mainly through abstract mathematics.' Still, we must use this uncommon sense sensibly, he advises.

As science evolves, so does our common sense, and they define each other as they evolve together. Remarks the late British theoretical physicist John Ziman in *Real Science: What It Is, and What It Means*: 'Thus, when we contrast a scientific belief with "common sense" we are indicating that we think we could define it precisely and give some coherent account of why it should be relied on, rather than just "taking it for granted" as an undeniable truth.'

Now, if proponents of extraterrestrial visitors move away from the common-sense belief of the impossibility of such beliefs, we may find common ground. This will be the new common sense and the new science of those unidentified 'alien' objects in our skies.

CHAPTER 2

MARTIAN MANIA

Smart Martians and silly scientists

Since Galileo caught sight of Mars through his newly built telescope in 1609, the red planet has been a source of intrigue and imagination. Until the mid-20th century, scientists seriously believed that intelligent life existed on Mars.

E.aT. Bell, the famous historian of mathematics, calls Carl Friedrich Gauss the 'Prince of Mathematicians' who can be ranked only with Archimedes and Newton, 'and it is not for ordinary mortals to attempt to range them in order of merit'. It has been said of Gauss that almost everything which the mathematics of the 19th century has brought forth in the way of original scientific ideas is concerned with his name. The question of life on other planets did not escape his attention either.

At that time, it was believed that Martians could look at Siberia. To attract their attention, Gauss proposed clearing stretches of the forest there to form a gigantic right-angled triangle with squares on each side, planting wheat in the triangle, and leaving squares of trees around it. He believed that his diagrammatic demonstration of Pythagoras' theorem would reveal our presence to our neighbours (how could Martian masterminds miss the mathematical message hidden in Pythagorean 'crop circles'?).

A few years later, Johann Joseph von Littrow, an Austrian physicist, worried about the poor visibility of Pythagoras' diagram in the dense Siberian forest, suggested that the right thing to do would be to dig giant ditches in the Sahara in the form of geometric figures such as triangles, squares and circles, fill them with water, and pour kerosene on top of the water and set them ablaze at night. Even this bright idea was not followed up.

The idea of a blazing Sahara as a beacon to attract Martians' attention appeared again in 1880 when French astronomer Camille Flammarion suggested that if chains of light were placed on the Sahara on a sufficiently generous scale to illustrate Pythagoras' theorem, intelligent Martians might conclude that there was intelligent life on Earth.

These ideas certainly fired up the imagination of many 20th-century scientists. In 1920 Guglielmo Marconi said that he had received mysterious radio signals that he believed might have come from Mars or some other region of the universe where electrons are in vibration. He ignored these signals, saying, 'I'm concerned enough at present with business upon Earth.' The *International Herald Tribune* reported: 'Mr. Thomas A. Edison, commenting on the statement of Marconi that untraced wireless calls might come from Mars, stated that such a thing is possible. 'Existing machinery is able to send signals to Mars,' said Mr Edison. 'The question is, have the beings there have instruments delicate enough to hear us? They say Martians are as far ahead of humans as we are ahead of chimpanzees. If that is true, they must have such apparatus.'

In 1924 David Todd, an American astronomer who, like Edison, believed in the superiority of Martian technology, convinced the US government to turn off high-powered radio transmitters on 23 August, when Mars came closest to Earth. On that day, when the two worlds were just 56 million kilometres (34 million miles) apart, transmitters were turned off for five minutes before each hour so that Todd could listen to Martian chatter. Martians were also aware of Earth's closest approach. They decided to turn off their radio stations and sit in silence on that day. They already knew that turning off relaxing music and sitting in silence is good for your health – something that our researchers found decades later.

Todd did not give up his hope of communicating with Martians that easily. He suggested filling a 15-metre (49-foot) bowl with mercury at the bottom of an abandoned Chilean mine and placing a powerful light source at the focus of this natural parabolic reflector. This time he did not convince authorities to act on his good advice.

Startling Martians with maths

Nineteen-twenty was a vintage year for scientific ideas about communicating with Mars. Pages and pages of *Scientific American* for that year are filled with suggestions not only for technical means of signalling but also about the 'language' of the message.

'Suppose we have decided to signal to Mars and have found a way to make our signals carry to Mars and decided what we want to say to Mars. What then?' asks a reader, identified only as W. L. B. from New York. 'Shall we signal in English, or French, or in what language? Perhaps the advocates of Esperanto would have us believe that it is such a simple and logical universal language that the Martians must have developed it and put it to use.

'Or perhaps the enthusiastic mathematicians who tell us, with delightful generality, that mathematics is the universal truth and that therefore the message will be such as to convey some fundamental mathematical idea, will deign to work out their scheme for us, and tell us with just what mathematical truth we are to startle the Martians?'

Someone suggested startling the Martians with an IQ test by sending two signals, then four, and if they answered back with eight signals, we would know that Martians could multiply.

Not everyone was convinced of the existence of Martians. 'Now, while it may be possible that Mars is inhabited by some sort of creatures, there could not be any possibility be like ourselves in any essential respect,' a reader, Hudson Maxim, wrote, 'They are just as likely to resemble straddle-bugs, spiders or ground-moles as they are to resemble us.' He continued that if all the letters of the Lord's prayer were thrown into the air, there would be as much chance of the letters 'falling back into their proper places to print the prayer without an error as there is of there being inhabitants on the planet Mars with whom we might by any possibility communicate'.

These views did not deter the editors of *Scientific American*, a weekly magazine then, from publishing a lengthy article, 'What shall we say to Mars?', on 'this rather unorthodox scientific matter' in the issue of 20 March 1920. The article described an elaborate scheme of writing a pictorial message (see figures) in 'dots or dashes, of wireless or light, as the case may be', and then using these pictures to teach Martians a new language. Image d) shows, directly beneath the picture, a group of dots and dashes. They spell, in Morse code, the word 'man'. The article suggested that the following messages be labelled 'hand' and 'head', and so on. This practice would compel the Martians to the conclusion that these marks were arbitrary symbols for the various objects pictured and thus would help them in 'building up the interplanetary language'. In practice, the article said, 'we shall not use English words, but some language of more logical construction, like Esperanto'.

The last serious proposition to attract the attention of Martians came in 1942 from the famous British astronomer James Jeans. He suggested shining a group of searchlights towards Mars and emitting successive flashes to represent a series of numbers. If, for instance, he said, the numbers 3, 5, 7, 13, 17, 19, 23 … (the sequence of prime numbers) were transmitted, the Martians might surely infer the existence of intelligent life on Earth.

The following figure shows how dots and dashes can be used to send pictorial messages to Mars. a) Dots and dashes arranged to form a simple diamond; b) Dashes of the previous message plotted on quadrilled paper; c) More elaborate message, with the dots again eliminated; d) A still more complicated message, depicting a man. (Reprinted with permission from *Scientific American*, 20 March 1920)

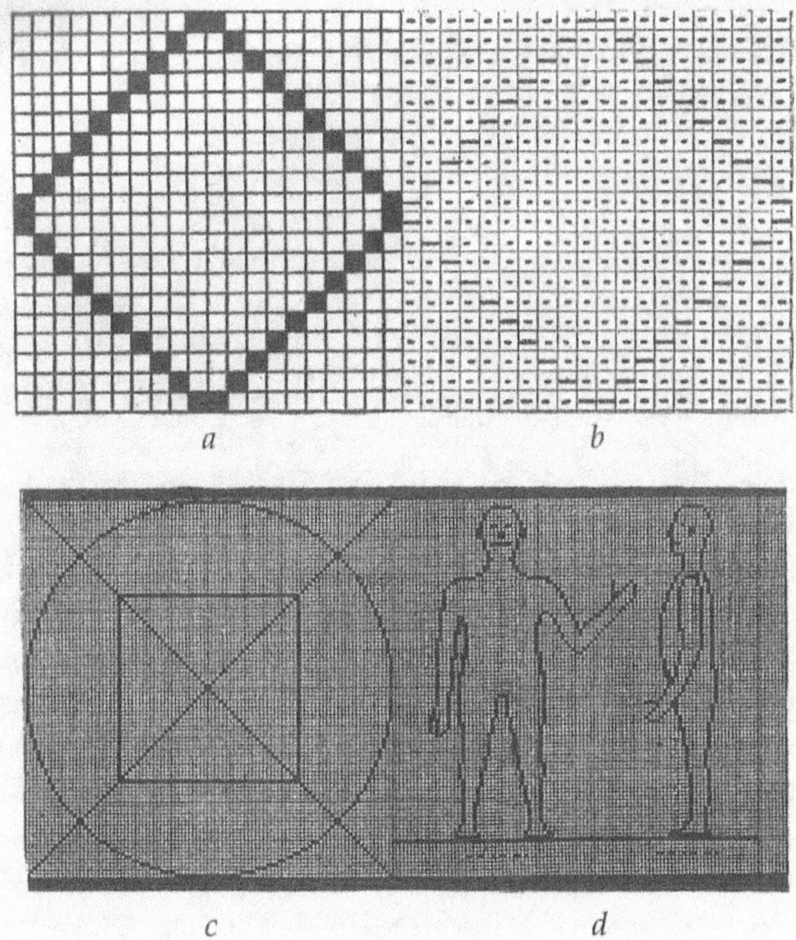

The day Martians invaded Earth

Yes, H.G. Wells' menacing Martians in *The War of the Worlds* did invade Earth to enslave humanity. At 8.50 pm on 30 October 1938, the day before Halloween, the marauding monsters (without their Halloween masks) arrived at Grovers Mill, a sleepy hamlet in New Jersey, at least on the radio:

> The metal casing is definitely extraterrestrial – not found on this earth … This thing is smooth and, as you can see, of cylindrical shape …

Someone's crawling out of the hollow top … I can see … two luminous disks – are they eyes? It might be a face. It might be … [SHOUT OF AWE FROM THE CROWD]

Good heavens, something's wriggling out of the shadow like a gray snake. Now it's another one, another one, and another one! They look like tentacles to me. I can see the thing's body now. It's large, large as a bear and it glistens like wet leather. But that face, it – ladies and gentlemen, it's indescribable. I can hardly force myself to keep looking at it, so awful. The eyes are black and gleam like a serpent. The mouth is V-shaped with saliva dripping from its rimless lips that seem to quiver and pulsate. The monster whatever it is can hardly move …

Ladies and gentlemen, I have a grave announcement to make. Incredible as it may seem … those strange beings who landed in the Jersey farmlands tonight are the vanguard of an invading army from the planet Mars.

Millions of listeners heard this 'live commentary' when they tuned in to a popular Sunday night programme on CBS radio, *The Mercury Theatre on the Air*, that featured plays directed by Orson Welles. The play that night was an adaptation of *The War of the Worlds*, presented as simulated news bulletins and scene broadcasts, presumably to heighten the dramatic effect. Welles had made another important change: he transplanted the action from H.G. Wells' Victorian England to contemporary America.

As the play unfolded, dance music from a hotel was interrupted several times by fake news flashes about a professor at an observatory noting a series of gas explosions on the planet Mars. Bulletins and scenes followed, reporting the landing of a 'meteor' in New Jersey, the discovery that the 'meteor' was a 'metal cylinder' containing strange creatures from Mars armed with 'death rays', and that nearly 7,000 soldiers had been crushed and trampled to death.

The radio drama was introduced as a work of fiction, but those who missed the announcement at the beginning (the next one didn't arrive until 40 minutes into the play) confused fiction for fact, and 'a wave of mass hysteria seized thousands of radio listeners throughout the nation between 8.15 and 9.30 o'clock,' the *New York Times* reported next day in a lead story – 'Radio listeners in panic, taking war drama as fact' – that filled nearly three-quarters of a page of the broadsheet. 'The broadcast, which disrupted households, interrupted religious services, created traffic jams and clogged communications systems, was made by Orson Welles, who as the radio character, "The Shadow," used to give "the creeps" to countless child listeners,' the report said. 'This time at least a score of adults required medical treatment for shock and hysteria.' Mass hysteria mounted so high that in some cases people told the police they saw the invasion.

The report recounted several examples of the panic, including: 'In Indianapolis a woman ran into a church screaming: "New York destroyed; it's the end of the world. You might as well as go home to die. I just heard it on the radio." Services were dismissed immediately.' And: 'The bartender of a tavern … closed the place, sending away six customers, in the middle of the broadcast to "rescue" his wife and two children.' The broadcast even fooled two Princeton University geologists: 'They armed themselves with the necessary equipment and set out to find a specimen. All they found was a group of sightseers searching like themselves for the meteor.'

The following figure shows the front page of the *New York Times*, 31 October 1938, with the lead story on a radio broadcast of a dramatisation of H. G. Wells' *The War of the Worlds*, which sent thousands of Americans fleeing their homes in panic.

Martian canals

The idea of intelligent life on Mars was, in fact, born in 1877 when Earth and Mars were in 'favourable opposition' – that is, their orbits had brought them to their closest – something which takes place every 30 years or so. Italian astronomer Giovanni Schiaparelli then noticed that the Martian surface was crisscrossed with a network of about 100 lines. He saw them again in 1879 and again in 1881. He called these lines *canali* ('channels'), which suggested that they were a natural feature. However, the Italian word was mistranslated into English as 'canals'. The idea of such large-scale artificial structures led to wild speculations about intelligent life on Mars.

However, the Italian word was mistranslated into English as 'canals'. The idea of such large-scale artificial structures led to wild speculations about life on Mars. The American astronomer Percival Lowell became the champion of those who believed that these canals were the work of intelligent beings on the red planet. His argument was based on the undeniably remarkable regularity of the canals. He pointed out that they are usually the shortest distances between the points they connect, and they meet in groups of three, five, seven, and more in well-defined spots. Lowell termed these spots 'oases', like many spokes converging in a wheel hub. To him, they were so artificial that they were the symbols of an intelligent civilisation that solved the vital problem of postponing the day when their planet might eventually dry up, and they perish.

In 1905 Lowell published a popular book about the Martian civilisation, which fuelled the public's fascination with Mars. (Carl Sagan once observed that intelligence was certainly involved in the creation of Martian canals. The only question was which end of the telescope the intelligence was on.)

Martian movies

When H.G. Wells wrote *The War of the Worlds* in the late 1890s (it was published in 1898), he was very much inspired by the works of Lowell and other astronomers.

He was also familiar with the historical debate about other worlds. The first edition of the novel has a prefatory quotation from the 17th-century German astronomer Johannes Kepler: 'But who shall dwell in these worlds if they be inhabited? … Are we or they Lord of the Worlds? … And how are all things made for man?'

At least after *The War of the Worlds*, we were the Lord of the Worlds: 'The Martians – *dead!* – slain by the putrefactive and disease bacteria … after all man's devices had failed, by the humblest things that God, in His wisdom, has put on this earth,' proclaimed the CBS radio drama. Incidentally, when the CBS drama was replayed in Ecuador in 1949, a riot broke out, resulting in the death of six people.

There are two film adaptations of the novel: Byron Haskin directed the 1953 original, Steven Spielberg directed the 2005 remake.

Mars has also inspired scores of other (almost all B) movies. A sampler: *Invaders from Mars*, 1953, 1986 (Martians brainwash residents of a small town); *Mars Needs Women*, 1968 (Martians invade Earth to capture women; almost proves the theory of panspermia, remarks a wit); *The Alpha Incident*, 1984 (deadly virus from Mars reaches Earth in a space probe); *Total Recall*, 1990 (Arnold Schwarzenegger travels to Mars); *My Favourite Martian*, 1999, adapted from the 1960s TV show (a friendly Martian visits Earth), *Mission to Mars*, 2000,(the first manned mission to Mars runs into a catastrophic disaster), *Last Days on Mars*, 2013 (another first mission to Mars meets tragedy), *The Martian,* 2015 (on yet another mission to Mars an astronaut is left behind by the crew) and *Settlers*, 2021 (refuges from Earth trying to survive the cosmic elements and each other).

Face on Mars: unmasked!

A century after 'the discovery of Martian canals', the speculations about life on Mars became much wilder when in 1976, NASA released a photograph taken by its unmanned Viking spacecraft, which was circling the red planet at a distance of 1,860 kilometres (1,162 miles), snapping photos of possible landing sites. The grainy image looked like a shadowy likeness of a human face. The caption was a bit of poetic license to engage the public and attract attention to its Mars missions:

'The huge rock formation ... which resembles a human head ... formed by shadows giving the illusion of eyes, nose and mouth.'

The Face on Mars soon became a pop icon and an urban myth that it was evidence of past intelligent life on Mars. In 1987 Richard C. Hoagland, an American science journalist, claimed in his book, *The Monuments of Mars: A City on the Edge of Forever*, that the face was the sculpture of a human being found near a ruined city whose 'forts' and 'pyramids' could still be seen. Like Stonehenge, the face pointed to where the Sun rose on the Martian solstice about 500,000 years ago when the gigantic face was constructed. Hoagland's book is still in print and propagating the myth.

NASA tells a different story. In 2001 the Mars Global Surveyor spacecraft captured the first high-resolution photo of the face. 'Each pixel in the 2001 image spans 1.56 meters, compared to 43 meters per pixel in the best 1976 Viking photo,' NASA claims. The picture now shows the Martian equivalent of landforms common around the American West. There are no eyes, no nose, and mouth on the 3-kilometre (nearly 2 miles) long and 240 metres wide (about 800 feet) high Face. It's the work of erosion on Martian rocks, not extraterrestrial beings.

But urban myths don't die quickly. A statement on the NASA website admits, 'Some people think the Face is *bona fide* evidence of life on Mars – evidence that NASA would rather hide, say, conspiracy theorists. Meanwhile, defenders of the NASA budget wish there was an ancient civilization on Mars.'

The first of the following two pictures show the grainy photograph taken by the *Viking* spacecraft in 1976, which started the Mars face controversy. The second picture is a high-resolution image of the so-called face on Mars taken by the *Mars Global Surveyor* spacecraft in 2001. (Photos courtesy of NASA)

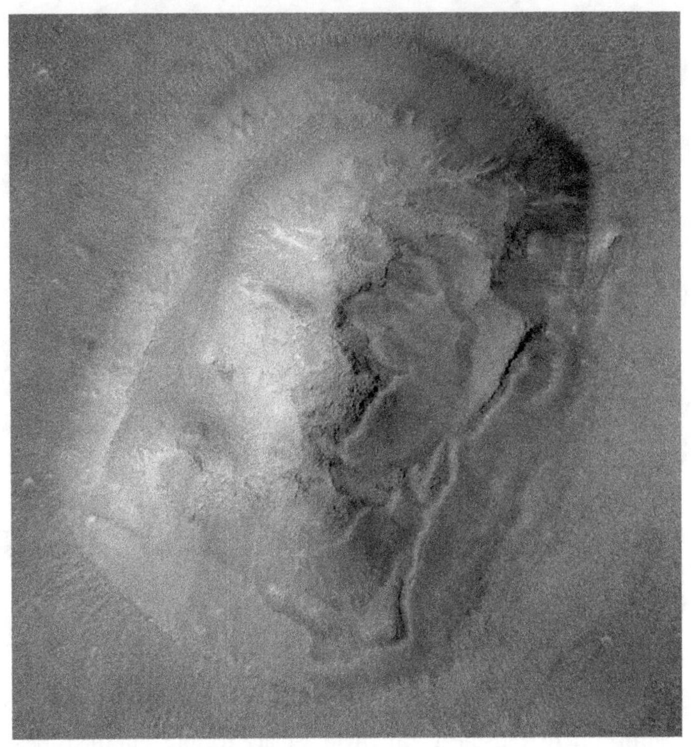

CHAPTER 3

OUR WORLDS

The eight planets

Outwards from the Sun:

- Mercury: Only slightly larger than Earth's moon; it is the smallest planet. With a furnace-like temperature, it is the hellhole of the solar system. No evidence of life.

- Venus: It seems like Earth's twin, but it is a parched and scorched world. It is hotter than Mercury because of the greenhouse effect as its dense atmosphere traps heat. An unlikely place for life, but microbes might exist high in the clouds where it is cooler. Scientists have found phosphine, a molecule that may have a biological origin, 50 kilometres (30 miles) above the surface.

- Earth: Slightly larger than nearby Venus, it is the fifth-largest planet. Liquid water, the elixir of carbon-based life, covers 75 per cent of its surface. No signs of life on its only moon.

- Mars: Life, not of the little-green-men variety, might have existed on this dusty, cold, red planet in the past, but the chances of life now are slim. It has two moons.

- Jupiter: One of its 79 known moons, Europa, fascinates astronomers more than the largest planet, as it may have a deep ocean beneath its icy surface. This gas giant is twice as massive as all the other planets.

- Saturn: Rainbows form after it rains on the ringed planet's largest of its 82 moons, Titan, but its 'water' is liquid methane. Geysers of water spew out of reservoirs just metres beneath the icy surface of its tiny moon, Enceladus, hinting at the possibility of life. Known as the jewel of the solar system, this gas giant has seven spectacular rings.

- Uranus: A blue-green wonder with 11 separate rings and 27 known moons. This icy planet has no life.

- Neptune: The smallest of the four 'gas giants', so-called because they do not have solid surfaces – the other three are Jupiter, Saturn and Uranus. It has 14 moons. Like Uranus, this icy planet has no life.

The dwarf planets. The most famous is Pluto (smaller than Earth's moon, the 'ninth planet' has now been relegated to this new class of objects); others are Ceres, Eris, Makemake and Haumea, and many more icy balls waiting to be classified.

The wanderers

Five wandering points of light in the night sky are easily visible to the naked eye and were known to the ancients – the Babylonians, the Egyptians, the Chinese, the Hindus and others. The Greeks called them *planetes* ('wanderers'). Today we know the five as Mercury, Venus, Mars, Jupiter and Saturn – all named after Roman gods and goddesses.

Even the inquiring Aristotle did not know that he was standing on the sixth planet – Earth. Earth, a Germanic word, is at least 1,000 years old. It simply means 'the ground'.

The ignorance that Earth is not a planet continued until 1543, when Copernicus declared that 'Earth was simply one wanderer of many'. In 1609, after learning of its invention in Holland, Galileo designed and built his own telescope. With his 20-power telescope, he was the first to see the wandering points of light in the night sky as spheres – as other worlds like Earth. In 1633 Galileo removed any remaining doubts about the motion of Earth from the minds of adherents of Aristotle when, after recanting his belief in the Copernican system at his trial by the Inquisition, he stamped his foot and muttered inaudibly: *Eppur si muove* ('And it does move.') Many doubt that such a protestation ever happened, but that kick, real or imagined, has kept Earth in perfect orbit ever since.

The Sun and its six planets were now a complete heavenly family, the solar system. No one, not even William Herschel, who discovered the seventh planet, Uranus, in 1781, imagined that there could be another planet beyond Saturn.

The first recorded discovery of a planet

Herschel was born in Germany but moved to England when he was 19 and lived there for the rest of his life. He was a talented musician and for many years worked as an organist at the Octagon Chapel in Bath. He became interested in astronomy when he was 35. A year later, in 1774, he saw the stars for the first time through a telescope, a telescope that he had built himself. He gradually gave up his musical career and became obsessed with constructing better telescopes and watching the sky through them. 'He was incessantly making fresh mirrors or trying new lenses, or combinations of lenses to act as eye-pieces or projecting alterations in the mounting by which the telescope was supported,' remarks Robert Ball in his 1895 rich biography of Herschel in *The Great Astronomers*. His younger sister Caroline, who at this time had come to live with him and look after the housekeeping, complained that in his 'astronomical ardour he sometimes omitted to take off, before going into his workshop, the beautiful lace ruffles which he wore while conducting a concert, and consequently they became soiled with the pitch employed in the polish of his mirrors'.

In the history of science, Herschel holds the position of the greatest astronomer of his time, Caroline, a grander position: the first important woman astronomer. Caroline not only managed household duties, but she also mastered the arduous art of polishing mirrors and the complex mathematics of astronomical calculations. When Herschel was at the telescope, from dusk till dawn, 'Caroline sat by him at her desk, pen in hand, ready to write down the notes of the observations as they fell from her brother's lips'. As the telescope was in the open air, during some winter nights, the ink would actually freeze in her pen. The next day she would carefully transcribe the observations during the night and make necessary calculations so that her brother could observe the next swathe of the sky.

On 13 March 1781, while systematically observing stars in the constellation of Gemini, William Herschel came across an object that appeared as more than a little point of light, as stars did in his 7-inch-aperture reflecting telescope. When he applied a higher magnifying power, the object appeared like a tiny yellow-green disc. He thought he had discovered a comet and reported it as such. He continued his observations and calculations for months, which showed that the object had a circular orbit, like a planet's, rather than the elliptical orbit of a comet. 'The organist at the Octagon Chapel at Bath had, therefore, discovered a new planet with his homemade telescope,' as Ball recounts the momentous, first recorded discovery of a planet in human history.

Herschel, an unknown astronomer until then, was now a celebrity. He named the planet 'Georgium Sidus' ('George's star') in honour of George III, King of England. The name was used in Britain for many years, but the planet was eventually named Uranus, after the father of Saturn in Roman mythology. King George gave Herschel and Caroline the titles of 'The King's Astronomer' and 'Assistant to the King's Astronomer' and life-long pensions. Herschel – and Caroline – continued to scan the skies until he died in 1822. Their discoveries include two moons of Uranus, two new moons of Saturn and numerous double stars, nebulae and clusters.

The controversy over the discovery of Neptune

The discovery of Uranus posed a problem for astronomers. They kept on finding it in the wrong parts of the sky; the planet was apparently drifting away from its predicted orbit. Was there an unknown planet beyond Uranus pulling it out of its orbit?

Urbain Le Verrier, a French mathematician, assuming that an unknown planet was the culprit, decided to calculate its size and position. He published his first results in June 1846. After more work, on 18 September, he wrote to Johann Galle at the Berlin Observatory to ask him to look for it. Galle received the letter on 23 September, and on the same night, he spotted a small blue disc remarkably close to the position where Le Verrier had predicted it. He wrote to Le Verrier on 25 September: 'Monsieur, the planet of which you indicated the position really exists … My God in heaven, it is a big fellow.'

On the other side of the Channel, *The Times* announced the discovery on 1 October with the headline: Discovery of Le Verrier's planet. It noted that some astronomers in London had also observed it on the evening of 30 September. The newspaper, however, did not know that the British also had a stake in the discovery. A young mathematician, John Couch Adams, also suspected that an undiscovered planet was affecting Uranus' orbit. In October 1845, he sent his predictions to the Astronomer Royal, George Airy, who wrote back in November asking a minor technical question. Airy did not show any enthusiasm for Adams' calculations until July 1846, when he heard Le Verrier's predictions. He met with James Challis, director of the Cambridge Observatory, and John Herschel, William's son and a prominent astronomer, and asked Challis to begin a search at the Observatory. However, he was doubtful of the success, as he wrote later: 'It was a novel thing to undertake observations in reliance upon merely theoretical deductions; and while much labour was certain, success appeared very doubtful.'

On 10 September, Herschel addressed a meeting of the British Association for the Advancement of Science in Southampton, in which he said that he saw the new planet as 'Columbus saw America from the shores of Spain. Its movements have been felt, trembling along the far-reaching line of our analysis.'

After reading the *Times* report, Airy prepared a set of documents in which he outlined Britain's case. Before these documents were published at the 13 November meeting of the Royal Astronomical Society, on 3 October, Herschel wrote to London's *Athenaeum* magazine making public the role of Adams. Support for the British claim also came from Wilhelm Struve, director of the observatory at Polkowa, near St Petersburg. He wrote in the *Athenaeum* (20 February 1847): 'But impartial history will, in future, make honourable mention also of the name of Mr

Adams, and recognise two individuals as having, independently of one another, discovered the planet beyond Uranus … But Mr Adams's labours were unsuccessful because the two astronomers (Mr Challis of Cambridge and Mr Airy of Greenwich) to whom they were known hesitated to admit them without further examination.'

French astronomers were sceptical of the British claim; nevertheless, they agreed to name Le Verrier and Adams the co-discoverers. British astronomers accepted the name, Neptune, after the Greek god of the sea, suggested by Le Verrier.

The following figure shows a cartoon by Amédée de Noé, a French caricaturist and lithographer who became famous as Cham, published in the French newsweekly *L'Illustration* (7 November 1846) at the height of the controversy over the discovery of Neptune. The cartoon shows a caricature of Adams looking through a telescope across the Channel at Le Verrier's notes.

Airy's account had always rankled some science historians and writers. In 1964, prolific science writer Isaac Asimov described him as 'a conceited, envious, small-minded man [who] ran the observatory like a petty tyrant'. In the mid-1960s, when historians looked for Airy's file at the Greenwich Observatory, they found it missing. It turned up in 1998 at the Chilean Institute of Astronomy. More documents have appeared since.

After analysing all the documents, a trio of international historians – William

Sheehan, Nicholas Kollerstrom and Craig Waff – concluded that though Adams did some interesting work, he never calculated the exact position of the new planet. In *Scientific American* (December 2004), they accuse Britain of stealing Neptune: 'Now the passions stirred by the international rivalries of the 1840s have died down … we can affirm that Adams does not deserve equal credit with Le Verrier for the discovery of Neptune … The achievement was Le Verrier's alone.'

After a century and a half, Neptune has been restored to its rightful discoverer, and it has completed only one orbit since its discovery: it takes Neptune about 165 years to orbit the Sun.

A self-taught astronomer finds the (ninth) planet

Even after Neptune's discovery, Uranus kept on misbehaving. The existence of Neptune had explained irregularities in its orbit – almost, but not quite. Percival Lowell, the wealthy American astronomer who had earlier championed the idea that intelligent Martians had created canals all over the surface of Mars, suggested that an undiscovered planet was affecting Uranus and Neptune. He called it Planet X (X as in unknown). In 1895 he started searching for it from the Lowell Observatory near Flagstaff in Arizona, which he had built a year earlier, partly for the very purpose of finding the new planet. The technique for the search for new planets has changed since the discovery of Neptune. Astronomers now made two photographs of the same part of the sky, several weeks apart. Using an optical device called a blink comparator, they then manually checked both photographs, made up of millions of dots, hoping to find a dot that had moved from its place from the earlier photograph (computers now do this painstaking task). Lowell spent 10 years searching through countless photographs, but he found nothing. A disappointed man, he died in 1916. However, his observatory continues to operate. In 1929 Clyde Tombaugh, an earnest high school graduate who was too poor to afford college, wrote to the Lowell Observatory, enclosing drawings of his observations of Mars and Jupiter through his homemade 9-inch-aperture telescope. Impressed with the quality of his work, the Observatory offered him a job working with their new 13-inch photographic telescope. After less than a year's searching and photographing the sky, on 18 February 1930, he saw a faint point of light moving a bit from night to night. The 24-year-old self-taught astronomer had found the ninth planet. It was named Pluto for the Greek god of the underworld (Mickey Mouse's dog and the new artificial element plutonium were subsequently named after the new planet).

What's a planet anyway?

What is really a planet? Dictionaries tell us that a planet is a large, round heavenly body that orbits a star and shines in the star's reflected light. This simple definition is good enough for you and me, but not for astronomers. According to the IAU, a planet must meet the following three criteria:

- The body must orbit the Sun. A tug from the sun's gravity causes the current eight planets to revolve around the fiery body along distinct paths.

- The body must have enough mass to become shaped into a sphere. As a planet spins around its axis, gravity's centre-pulling force moulds the planet into a sphere. The greater an object's mass, the greater its gravity, and the more likely it will form into a sphere.

- The body's path around the Sun must be clear of other celestial bodies. Scientists think that as a planet forms, its gravity pulls in the surrounding material, such as dust and gas, causing the planet to grow.

Pluto does not meet the third criteria. As a resident of the Kuiper belt, it occupies the same area of space many other large rocky bodies. The Kuiper belt is a region of the solar system beyond the orbit of Neptune, which has many comets, asteroids and other small icy bodies.

Not everyone is happy with Pluto's demotion to a dwarf planet. Some astronomers prefer a 'geophysical definition' of a planet: A planet is a sub-stellar mass body that has never undergone nuclear fusion and has sufficient self-gravitation to assume a spherical shape adequately described by a triaxial ellipsoid regardless of its orbital parameters.

Don't waste time trying to comprehend this new definition. But if accepted, it will allow you to go back to the mnemonic you learned at school to list the order of nine planets (outwards from the Sun): My Very Educated Mother Just Served Us Nine Pizzas. The demotion of Plato had left your mother to serve nine nothings.

Planet X: the saga of a hypothetical planet

After the discovery of Pluto in 1930, astronomers asked 'is it the mysterious Planet X?' No. Pluto is so tiny that it has no appreciable effect on the motion of either Uranus or Neptune. In the 1990s, astronomers recalculated the masses of Jupiter, Saturn, Uranus and Neptune, using data from Voyager fly-bys, and concluded that Uranus and Neptune were right on course. No Planet X is affecting Uranus and Neptune.

Pluto has now been relegated to a new class of objects called dwarf planets. This new category also includes Ceres, discovered in 1801. In 2003 American astronomers discovered an object they thought was Planet X. They informally called it Xena as a wordplay on Planet X (in 2006, it was officially named Eris). Like Pluto, Eris is a rocky ball covered with frozen methane. A little larger than Pluto, it lies about three times as far from the Sun as Pluto, making it the furthest object ever seen in the solar system. It has also been classified as a dwarf planet.

But the myth of Planet X refuses to die. Conspiracy theorists say that the orbit of this mysterious and rogue planet is coming closer to Earth and causing extreme weather and major earthquakes. They also blame it for an increase in solar flares erupting from the surface of the Sun. Solar flares rise and fade slowly over the years. When a solar flare peaks, it unleashes waves of energy, disrupting Earth's magnetic field, which in turn can damage power grids and communication satellites. Our dependence on technology places us at greater risk if power grids and satellites stop working.

In 2016, California Institute of Technology astronomers Konstantin Batygin and Mike Brown showed by mathematical modelling – not direct observations – that there is evidence of a giant planet tracing an unusual, elongated orbit in the out solar system. They believe this so-called planet X (in fact, it would be 'Planet Nine' if discovery is confirmed) has a mass about 10 times that of Earth and would take it between 10,000 and 20,000 years just to make one full orbit around the Sun (the farthest planet Neptune completes an orbit roughly every 1565 years).

The anthropic principle

There is nothing special about the conditions on our planet. There is no reason things should be different anywhere else in the universe. In other words, one's location is unlikely to be special, or to put it bluntly, wherever or whenever we are, it is nothing special. This is known as the 'Copernicus principle' after Copernicus, who in 1543 demoted Earth to an ordinary unprivileged place in the cosmos.

In contrast, the anthropic principle (from Greek *anthropos*, 'human beings') maintains that humans hold a special place in the universe. The fundamental laws of physics that govern the universe are not the result of chance but somehow fine-tuned to allow the existence of intelligent life. If, for example, the force of gravity was slightly different from what it is now, there would be no Sun-like stars anywhere.

Simply put, we live in a Goldilocks universe ('Ahhh, this porridge is just right') because the universe has 'just right' conditions for the existence of life. The universe must be the way it is; it was made for us and only us.

The term 'anthropic principle' was proposed by Brandon Carter, a British cosmologist, in 1973 at a symposium in Poland commemorating Copernicus's 500th birthday. He suggested two versions of the principle: 1) the weak anthropic principle: the conditions in the universe are compatible with our existence; and 2) the strong anthropic principle: the universe must have those properties which would allow intelligent life within it at some stage. Other scientists have suggested different possible implications of the strong principle, including that the universe was 'designed' to sustain human beings. This has been interpreted as evidence of a creator.

However, many cosmologists dismiss the anthropic principle as non-science because it makes no testable predictions. The celebrated British cosmologist Stephen Hawking and Leonard Mlodinow write in *The Grand Design*: 'The idea that the universe was designed to accommodate humankind appears in theologies and mythologies dating from thousands of years ago ... That is not the answer to modern science. As recent advances in cosmology suggest, the laws of gravity and quantum theory allow universes to appear simultaneously from nothing. Spontaneous creation is why there is something rather than nothing, why the universe exists, and why we exist. It is unnecessary to invoke God to light the blue touch paper and set the universe going.'

Hawking and Mlodinow also say that our universe seems one of many, each with different laws. 'The idea of a multiverse is a consequence predicted by many theories in modern cosmology. If true, it reduces the strong anthropic principle to the weak one, putting the fine tunings of physical law on the same footing as the environmental factors.'

This leads to only one conclusion: the universe is just one of many. It's not custom made for us.

Geocentricity: neither science nor religion

Geocentrism is the fiction that Earth is the centre of the universe as opposed to the fact of heliocentrism that Earth and planets revolve around the stationary Sun.

Amazingly, there are still people who reject well-established scientific theories and believe that Earth does not move at all. The modern geocentrism began in 1967 when Walter van der Kamp, a Dutch Canadian school teacher, circulated a paper to some Christian individuals and organisations arguing the case for the doctrine that Earth is the centre of the creation. In 1971 Kamp founded the Tychonian Society, which was named after the famous 16th-century Danish astronomer Tycho Brahe who dissented from the Copernican theory and accepted without question the dogma that Earth stood still at the centre of the universe and the Sun went around it ('The Sun', he declared, 'is the Leader and the King who regulates the whole harmony of planetary dance'). *The Bulletin of the Tychonian Society* dedicated itself to defending a geocentric universe and featured articles on the history, philosophy and scientific arguments for geocentrism.

In 1991 the Tychonian Society was reorganised as the Association for Biblical Astronomy, and its bulletin became *The Biblical Astronomer*. The Official Geocentricity Website claims: 'Of all the sciences, the Holy Bible has more to say about astronomy than any other ... This site is devoted to the historical relationship between the Bible and astronomy. It assumes that whenever the two are at variance, it is always astronomy ... that is wrong. History bears consistent witness to the truth of that stance.'

The basis of modern geocentrism belief is the literal interpretation of the Bible. They may consider themselves the standard-bearers for an ill-conceived fusion of science and religion. Still, science doesn't seek evidence for its theories in religion, and religion doesn't need science's approval for its beliefs.

The planets of our lives

The sky was once full of portents. A comet or conjunction of planets filled ordinary people with terror. The perception of the influence of planets on human lives could make mighty emperors and generals look up to the heavens for guidance. Since Isaac Newton's gravity forced order on the wandering of planets and other heavenly bodies, astrology has been intellectually weightless. Yet, it has not lost its seductive powers to charm us.

Though astrology and astronomy had a starry relationship in the past, astrology is not a science. Obviously, astrologers disagree. They even assign astronomy a lower status as an offspring of astrology. Astronomy has made remarkable progress in the past few centuries, but astrology has hardly changed. Tell this to astrologers, and they would cast the 'horoscope of their 'science': astrology is truth, and truth never changes. But we know now that some ancient 'truths' are not eternal. For example, when astrologers say that the Sun is in Aries between 21 March and 21

April, they mean that the Sun, as seen from Earth, is in the same part of the sky as the constellation of Aries. They ignore the astronomical phenomenon of precession, the slow, conical motion of the Earth's axis of rotation which makes Earth wobble around its axis in a 25,800-year cycle. Precession makes Earth's position compared to the constellations of the zodiac change over centuries. The Sun is no longer in the constellation of Aries; it's in the constellation of Pisces, and in 2597 it will move into Aquarius. Shouldn't it affect you if you are an Aries or any other astrological sign?

The ancients believed that Earth was the centre of the universe, fixed and unmoving. They saw the planets generally moving relative to the fixed stars from west to east. But occasionally, a planet will slow down, stop and then reverse direction (retrograde). They explained this puzzling motion by saying that the gods were annoyed and therefore 'malefic'. How could the 'malefic' or other positions of planets influence our lives?

Five centuries have passed since Nicolaus Copernicus displaced Earth from its throne and set it in motion. Astronomers have now discovered thousands of planets beyond the solar system. In comparison, the planets of astrology are indeed puny, and their influence on Earth insignificant.

Scientists talk about four forces that influence everything in the universe. Gravity is the long-range force: it holds chairs to the floor and planets in their orbits. The electromagnetic force is the attraction and repulsion between charged particles: it enables light bulbs to glow and lifts to rise. The strong force keeps atomic nuclei together: it binds the protons and neutrons in an atomic nucleus. The weak force is also a nuclear force: it causes elementary particles to shoot out of the atomic nucleus during the nuclear decay of such radioactive elements as uranium. The strengths of the forces vary widely. The strong force is one sextillion (1 with 36 zeros after it) times stronger than gravity, making gravity the weakest force. The strong force may be strong, but its reach is limited to the atomic nucleus.

Gravity influences us, but the tug of distant planets on you is smaller than the gravitational pull of this book. Carl Sagan once pointed out that the gravitational pull of the midwife would have a far more significant influence at a child's birth than the pull of a planet.

Does astrology work? Many significant studies have tried to answer this question, but none has found any proof that the astrological sign you were born under influences the person you become. The largest test of astrology ever taken – a statistical analysis of the birthdays of more than 20 million married people in England and Wales – has shown that their astrological sign has no impact on the probability of marrying and staying married to someone of any other sign. There is no such thing as zodiac signs and love compatibility. Googling 'love sign astrology' would produce hundreds of thousands of hits, but none of them will make you wiser in choosing your life partner.

Many studies, however, have shown that being born at certain times of year relates to a small but significantly increased risk of problems such as depression, schizophrenia and anorexia nervosa. People born in late winter and early spring in the northern hemisphere have a slightly higher risk of developing schizophrenia than those born at other times of the year. Similarly, northern hemisphere autumn birthdays are associated with an increased likelihood of suffering panic attacks. These are medical studies trying to find out how seasons affect mental health; in no way do they support astrology.

As a science, astrology is silly; as entertainment, it's harmless.

CHAPTER 4

OTHER WORLDS

Planets around faraway stars

Ever since the 18th century, when German philosopher Immanuel Kant and French mathematician Pierre Simon de Laplace put forward their nebular hypothesis for the formation of the solar system – a spinning cloud of gas and dust broke into rings that condensed to form the Sun and the planets – the possibility of planets orbiting other stars has fascinated scientists and science-fiction writers alike. The writers simply let their spaceship travel faster than light to exotic planets in the Milky Way, but scientists had to find a planet before even dreaming of travelling to it.

The search for exoplanets (or extra-solar planets) started in earnest in 1981 when a team of Canadian astronomers led by Gordon Walker observed dozens of Sun-like stars. Their decade-long search yielded nothing. It appeared that, after all, we are alone. At least it did until October 1995, when the news of the confirmation of a planet orbiting another star hit the headlines around the world (A planet orbiting a sunlike star challenges notions that earth is unique – *New York Times*).

The planet, discovered by Michel Mayor and Didier Queloz of Geneva University, orbits 51 Pegasi, a Sun-like star only 44 light-years from the Sun. It's a Jupiter-like gas giant 140 times the mass of Earth and so close to its star that it whips around it in a mere four days.

It seems that Mayor and Queloz opened the floodgates. Astronomers have since discovered thousands of exoplanets, many of them in multi-planet systems. Kepler Space Telescope, launched in 2009 and operated until 2018, helped detect many hundreds of exoplanets. Another space telescope, Transiting Exoplanet

Survey (TESS), launched in 2018, has already detected thousands of exoplanets, and more are being found every day.

Discovering exoplanets, especially smaller planets, is extremely difficult. We can rarely see an exoplanet directly. Only a handful of exoplanets have been found by seeing them through a telescope, or direct imaging as astronomers call it. These are gas giant planets orbiting very far from their stars.

Most exoplanets are found through indirect methods. Planets reflect light from the stars they orbit. The reflected light is millions of times dimmer than the light from the star. As planets are close to their star, it's difficult to distinguish this dim light from the star's bright light. Planet hunters use various indirect techniques.

One of them is the 'wobble technique'. Here's how it works. A planet orbiting around a star is like two whirling dancers pulling each other. As a planet circles its star, its gravitational pull on the star causes the star to wobble back and forth. The amount of this tiny wobble depends on the mass of the planet, the heavier the planet, the more the star wobbles. This wobble creates a 'Doppler shift' in the spectrum of the star as seen from Earth.

If you have noticed the changing pitch of the whistle of a train as it runs past you standing on a platform, you know what the Doppler shift is. The pitch is higher when the train is approaching you, and lower when moving away from you. The effect was described in 1842 by Austrian physicist Christian Doppler and is caused by the change in frequency of a sound or light wave with reference to the observer as the source moves away from or towards the observer. Three years later, Dutch meteorologist Christoph Buy Ballot tested the effect of sound waves in an endearing experiment – as a moving source of the sound, he used an orchestra of trumpeters standing in the open car of a railroad train, whizzing through the Dutch countryside. In 1848, French physicist Armand Fizeau showed that the Doppler shift applies to light coming from distant stars. No, he didn't use a spaceship full of trumpeters to test his idea. If a star is moving away from us, the light it emits shifts slightly towards the red end of the spectrum, making it appear slightly redder. If a star is moving closer to us, it appears slightly bluer. This phenomenon is called the redshift.

The spectrum of the wobbling star can reveal details such as the mass and the orbit of the planet, but not its nature. The wobble technique works well for massive Jupiter-like planets, but it fails to detect lighter Earth-like planets.

Another planet-hunting technique is gravitational microlensing, which can pick up smaller planets. Microlensing is based on Einstein's theory of general relativity, which says that the mass of an object causes space to curve. As a result, light can be bent by gravity. When an object passes between a star and Earth, its gravity causes the light to bend towards Earth. The light is magnified, or 'lensed', and the star seems brighter for a few hours, days or weeks. This spark of brightness makes planet-hunters beam from ear to ear; it is the tell-tale sign of the presence of a planet.

What makes a planet fit for life?

The planet must not be too hot, nor too cold. Its distance from its star determines whether it will be hot or cold. It must be at the right distance from its star for the existence of the elixir of carbon-based life – liquid water. It must not be too big, nor too small. If it is too big, its gravity will attract gases from space. It will then be like Jupiter, which has an outer shell of hydrogen and helium. If it is too small, like the Moon, its pull of gravity will be so weak that it will not be able to hold on to oceans and an atmosphere, as water or gases will be lost in space. It must also be relatively safe from space hazards: bombardment by asteroids and comets and blasts of dangerous radiation from space.

For an intergalactic estate agent our planet is prime real estate. It has the three 'must have' things in real estate: location, location and location. 'Any extraterrestrial civilization seeking a new world would place our solar system on their home-shopping list,' informed *Scientific American* in 2001. Pray that old copies of *Scientific American* are not found in bug-eyed alien doctors' consulting rooms.

Earth not only has the right location, but it also has a solid surface that can support water. This surface has a unique feature – the movement of the planetary crust across the planet's surface. Called plate tectonics, it provides many other life-supporting features. The Earth's crust is like a cracked eggshell. It is broken up into six large pieces and several smaller ones. The plates – as the pieces are called – are 100 to 150 kilometres thick and carry the continents and oceans on their backs like giant rafts. The plates are drifting about 2 centimetres per year (the rate at which our fingernails grow) over a layer of red-hot melted rocks that underlie the rigid crust. If these drifting plates collide, they cause mountains to rise. If they move away from each other, new oceans are formed. These changes take place over millions of years. Most volcanoes and mountain ranges are concentrated along the plate boundaries (or where the boundaries used to be). No other 'terrestrial' planet has a linear mountain chain or shows evidence of plate tectonics. However, magnetic maps of Mars suggest that plate tectonics may have occurred on the red planet in its early history.

Atmospheric gases, mainly carbon dioxide and water vapour, form a blanket around Earth and keep it at a comfortable temperature for life. Without this blanket, Earth's temperature would be below freezing point. Volcanoes are critical in releasing carbon dioxide into the atmosphere. Plate tectonics recycles chemicals crucial to keeping the level of carbon dioxide in the atmosphere uniform. It also makes possible the Earth's magnetic field, which shields us from the solar wind (a spray of charged particles from the Sun) and other lethal cosmic radiation. If plate tectonics stopped today, in a few million years, the mountains would erode, and our world would return to its state of 4 billion years ago:-the entire planet covered by a global ocean, except for an odd volcano poking out.

The simplest life forms take in water, carbon dioxide and nitrogen as nutrients and then release oxygen into the atmosphere. If we can find a planet with water, carbon dioxide and oxygen (or ozone, a form of oxygen usually found in the upper atmosphere), we will be one step closer to finding life – not necessarily complex life – on other worlds.

Goldilocks zones

Where can we find a planet fit for life? In a habitable zone, of course. A habitable zone, also known as a Goldilocks zone ('Ahhh, this porridge is just right'), is a region around a star in which a planet's surface temperature is 'just right' for liquid water to exist. The Sun's habitable zone extends from after Venus to just before Mars, so only our planet is within this zone. These are the present limits of the zone. But during the early history of the Sun, when it emitted less energy, Venus and Mars were probably also within the larger habitable zone. Perhaps water existed on these two planets in their early history. As the Sun heated up, Venus, the planet with a thick carbon dioxide and sulphur dioxide atmosphere and closer to the Sun, also heated up and lost its ability to hold liquid water. However, the geological history of Mars, which has very little atmospheric carbon dioxide, shows that it actually cooled off and also lost its ability to hold liquid water. Only the planet in the middle, with its correct balance of greenhouse gases, remained 'just right' like Goldilocks' porridge.

Europa, Jupiter's second moon, is an enigmatic world that defies the concept of a habitable zone. It has a deep ocean of liquid water under its icy crust. For us earthlings, water equals life. NASA scientists believe that if there is life on Europa, it almost certainly was completely independent of the origin of life on Earth. It would mean the source of life must be pretty easy throughout the universe. Is Europa's ocean teeming with aliens? Stay tuned. In 2024, NASA's Clipper spacecraft will soon make multiple flybys of Europa to investigate the possibility of life.

Our solar system is, of course, within the galactic habitable zone, but how far does the zone extend? A team of astronomers led by Charles Lineweaver of the University of New South Wales has identified a ring that measures 21,000 to 27,000 light-years from the galaxy's centre as the galactic habitable zone. This region contains about 10 per cent of the galaxy's 250 billion stars, which were formed between 4 and 6 billion years ago. Because a parent star and its planets were formed from the same cloud of gas and dust, the star must have the right 'metallicity'; that is, enough heavy elements (elements heavier than hydrogen and helium) to sustain life.

The right metallicity is not the only prerequisite for life. The star must have lived through relative cosmic calm – not too many supernova explosions – for at least 4 billion years, the time it took for life to evolve on Earth. This prerequisite rules out dangerous inner zones of the galaxy from the habitable zone. 'If there are aliens, 75 per cent of them will have had a longer time to evolve than we have,' Lineweaver says. 'This may be the most fundamental take-home message of this study.'

American astrophysicist William Danchi and French colleagues Bruno Lopez and Jean Schneider suggest that planet-hunters should look for planets around dying red giant stars. In a few billion years, when our Sun becomes a red giant star, its habitable zone will lie beyond Uranus. Planets that are at present very cold and icy would warm up and could kick-start life. The researchers' analysis shows that the period over which these conditions change is very long – long enough for life to form. It took about 700 million years for life to emerge on Earth. They suggest that meteorites from a planet could transport microorganisms from where life is ending to a planet where favourable conditions for its re-birth are emerging.

There are nearly 150 sub-giant and giant red stars within 100 light-years of Earth (compared to about 1,000 Sun-like stars). It narrows down the search area for planet-hunters. As the search for other planets is also, in part, a search for life beyond Earth, we may be closer to the holy grail of astrobiology, the study of the likelihood of extraterrestrial life.

This story of brave new worlds has a punchline, and it comes from Freeman Dyson, who believes in the possibility that life elsewhere is already adapted to living in a vacuum. Asteroids and comets, where gravity is weaker and moving from one world to another is easier, are the most likely habitats for life adapted to a vacuum. A planet would be a death trap for such forms of life, he says. 'A planet for them would be like a deep well full of water for a human child.' He suggests that the swarm of smaller objects in the Kuiper belt in our solar system – and Kuiper belts in other star systems – have much more real estate, which provides more habitable surface area than all planets put together. 'I am willing to bet even money that when the first alien life is found, it will not be on a planet,' he says. 'This is a bet which I will be happy either to win or lose.'

Before you dig into your pocket to place your bet, let's look at what other scientists have to say about the search for Earth 2.0

Earth 2.0

Astronomers predict that there could be as many as five billion habitable planets in our galaxy. Could one of them be like present-day Earth, a blue world covered with water and surrounded by clouds of white? The discovery of an Earth-like planet, Earth 2.0, will be a momentous moment in history.

Red dwarfs, the most common types of stars in our galaxy, are a good place to search for Earth-like planets. There is one catch: these stars pump out deadly amounts of radiation on their planets and thus frying chances of life as we know it.

Another type of stars, known as M dwarfs, is the smallest and coolest stars. They are smaller and less bright than the Sun, but they are 10 time as numerous. Again, there is a catch: M dwarfs take hundreds of millions of years to develop from gas clouds. Planets forming around these protostars would cook in the protostars heat and light and might boil off their life-giving water.

'The term "Earth-like" also carries some burdensome baggage,' warns NASA scientist Pat Brennan. 'That begins with how we define it. How could we recognise these qualities on a planet hundreds or thousands of light-years away?'

The truth is that if we do discover Earth 2.0 it could be vastly different from the Earth we live on today.

CHAPTER 5

LIFE AS WE KNOW IT

The secret of life

In 1944 Erwin Schrödinger, the celebrated physicist famous for his 'cat' (not a real moggy, but a thought experiment), wrote a little book, *What Is Life?*. Stepping outside his field of expertise, he speculated that life's genetic information had to be compact enough to be stored in molecules in 'some kind of code-script'. These molecules, passed from parent to child, are 'the material carrier of life'.

Nobel Laureates Francis Crick and James Watson, crackers of life's code, both have acknowledged how this revolutionary idea inspired them. In 1953, in 'a few weeks of frenzied inspiration', as *Time* magazine put it, the two young and unknown scientists solved the secret of life at the Cavendish Laboratory in Cambridge. On 28 February, Crick walked into the Eagle pub in Cambridge and, as Watson later recalled, announced that 'we have found the secret of life'. That morning they had discovered the last piece of the puzzle that revealed the double helix structure of DNA. In short, DNA consists of a double helix of two strands coiled around each other. When the strands are uncoiled, they can produce two copies of the original. This unique structure explains how DNA stores genetic information and how it passes this information on to the next generation by making an identical copy of itself.

The DNA (deoxyribonucleic acid) molecule is like a twisted ladder. Each 'side of the ladder' is made up of chains of alternating sugar and phosphate units. 'Rungs' are made from pairs of four chemical compounds called bases: adenine (A), thymine (T), cytosine (C) and guanine (G). The bases always pair in a specific manner: A pairs with T, and C pairs with G. Thus, there are only four combinations: A–T, C–G, T–A and G–C. The genetic code is the sequence of bases along the length of DNA. This code determines the order in which amino acids are linked together to form proteins.

The DNA resides in the nucleus of the cell. It instructs the cell to make proteins that control all the chemical processes in the cell. It does it by making RNA (ribonucleic acid), a close cousin of DNA. RNA is made up of the same bases as DNA, except that the base U (uracil) replaces the base T. RNA serves as the blueprint for making proteins needed by the cell. The instructions on RNA are in the form of a code that consists of combinations of three bases available on DNA. Each combination represents an amino acid. Of the 64 combinations possible, 61 represent the 20 amino acids (as only 20 amino acids occur in the cells of all organisms alive today, in some cases, several combinations refer to the same amino acid). The other three combinations act as 'full stops' in the coded information. The code can be written in either DNA triplets or the RNA copy of triplets. As an example, the triplet TTA in DNA (or CUU in RNA) instructs the cell to add the amino acid leucine.

The sequence of base pairs along the length of the strands is not the same in the DNAs of different organisms. This difference in the sequence makes one life form different from another. The human genome, the full DNA sequence of humans, has about 2.9 billion base pairs, which are wound into 24 distinct sausage-like bundles, or chromosomes. A gene is a segment of a chromosome. It is a length of DNA which has a complete code for one protein. All life forms have DNA.

What really is life?

Although it is difficult to define life, everyone agrees that all living things must have a system for storing and duplicating instructions about their structure. For life on Earth, DNA provides such a system. Its most remarkable feature is its genetic code, which is the same for all life forms that exist on this planet. And this feature makes the genetic code as old as life itself.

Life is a self-sustaining chemical system capable of Darwinian evolution. This is how NASA's Astrobiology Institute defines life. Charles Darwin presented his theory of evolution in his monumental book, *On the Origin of Species*, published in 1859. He said, in brief, that all present-day species have evolved from simpler forms of life through a process of natural selection. Advances in modern biology, especially in the knowledge of DNA, have enriched the theory of evolution. The modern view of evolution is still based on the Darwinian foundation: evolution through natural selection is opportunistic, and it takes place steadily.'

There are several hundred more definitions of life other than the above definition. So far, we have only one example of life, that is, life on our planet. 'Some astrobiologists argue that we really won't know what life is until we find an alternative to the basic structure found on Earth – the same DNA, metabolism and carbon blueprint shared by all known life, ET could tell us what life really is,' says Marc Kaufman of NASA's Astrobiology Institute.

Carbon copies of the last universal common ancestor

Everything alive is made up of cells (living cells were first described in 1665 by the English scientist Robert Hooke, who named them so because they reminded him of tiny monks' rooms in monasteries). Some bacteria have only one cell; some animals have millions.

The last universal common ancestor (LUCA) of modern cells, from which all organisms descended, lived about 3.8 billion years ago. Although LUCA was not necessarily the first living organism, hyperthermophiles, extremophiles that live at temperatures over 80 degrees Celsius in deep-sea hydrothermal vents, are probably its closest living relatives. When LUCA was born, the planet's atmosphere lacked oxygen, and abundant minerals in hydrothermal vents provided the energy needed to sustain life. Hyperthermophiles are the most ancient living forms known. Even if they are LUCA's closest relatives, it does not necessarily follow, as claimed by Gold, that life itself originated in deep-sea hydrothermal vents.

As there are no fossil records of LUCA, and no other known 'footprints',

LUCA's identity has proved elusive. Molecular biologists are trying to create a portrait of the mother of life by comparing the genes of all life forms. Even this method has flaws. Biologists estimate that a simple LUCA might have had as few as 450 genes, and its diameter was about 300 nanometres. A recent study of genomes of 100 species has found only 60 genes in common. Apparently, genes have been lost from species' genomes as organisms adapted to new conditions.

While scientists are still searching for LUCA's identity, they are sure that it has three things that are present in the cells of all living organisms today: DNA for genetic code, RNA for retrieving the genetic code for making amino acids, and ribosomes, molecular factories for assembling proteins from amino acids.

Here is a breakdown of the components of cells and the chemical elements present in them: *water*, universal solvent (hydrogen, oxygen); *protein*, involved in all functions (carbon, hydrogen, oxygen, nitrogen, sulphur); *fat*, energy storage (carbon, hydrogen); *carbohydrate*, cell walls (carbon, hydrogen, oxygen); *DNA*, *RNA* and a tiny molecule called *ATP*, genes and energy (carbon, hydrogen, oxygen, nitrogen, phosphorus).

As far as elements are concerned, cells consist mostly of carbon, hydrogen, oxygen and nitrogen, together with small quantities of sulphur and phosphorous. Plants and animals also require numerous other trace elements (small concentrations of elements present in a sample) such as sodium, potassium, magnesium, iron, zinc, calcium, chlorine, fluorine and copper. Of the 92 elements present in nature, only 21 play a role in life.

Carbon, hydrogen, oxygen and nitrogen are the prime elements of life. Why is life made of these elements and not others? The answer is simple: because they are four of the five most abundant elements on Earth. The fifth is helium, but it's an inert gas and takes absolutely no part in any known chemical reactions.

The story of life as we know it is the story of water and carbon. Without them, there would be no life on our planet, but could there be life without water or carbon (or both) on other planets? That we do not know – yet.

Water is a miracle that makes our world possible. It has many amazing characteristics that help to support and sustain life:

Supersolvent: The ability of water to dissolve almost everything enables it to carry nutrients through the bodies of plants and animals.

Ability to climb against gravity: Water can move easily through extremely narrow tubes and ooze through invisibly tiny holes. This 'capillary action' lifts water up from under the ground, through the soil to the roots of plants. It then ascends through stems and leaves.

A hearty appetite for heat: Water can absorb more heat than any other substance without a considerable rise in temperature. The slow cooling and warming of water prevent extreme climatic changes and protects living things from the shock of abrupt temperature changes.

Ice is lighter than water: Practically every substance contracts as it becomes colder, but water – when cooled below 4 degrees Celsius – expands to have about 10 per cent more volume as a solid than as a liquid. If ice did not float, oceans and bodies of water would be frozen from the bottom up, and there would be no living things in them.

The peculiar behaviour of water results from the so-called hydrogen bond between molecules of water. The hydrogen bond links the positively charged nucleus of a hydrogen atom in one water molecule to the negatively charged electron cloud of a nearby oxygen atom in another water molecule. There is another property of water that makes it unique. It is a polar molecule, which simply means that one part of the molecule has a positive charge and the other a negative charge. Other polar molecules can dissolve in water, but non-polar molecules cannot. The walls of cells are made of non-polar carbohydrates, ensuring that they will not dissolve in water.

Some scientists have suggested that liquids such as ammonia or methane – Saturn's moon Titan has oceans of methane – could replace the water in life forms that evolved on other planets. Two- to four-centimetre-long centipede-like worms have been found living in mushroom-shaped mounds of frozen methane seeping up from the floor of the Gulf of Mexico.

Life began when organic molecules, molecules having carbon, slowly began to assemble in liquid water. Carbon has an extraordinary ability to form compounds with other elements. A carbon atom can form bonds with four other atoms. These atoms may be other carbon atoms or non-metal atoms, especially hydrogen, oxygen, nitrogen, sulphur and phosphorus. Most organic compounds are combinations of carbon with one or more of these five atoms. Carbon atoms link together to form straight chains or rings of usually five to eight atoms. Some organic molecules contain as many as 100,000 atoms.

Carbon-based molecules in terrestrial life have two limitations: a) they cannot obtain liquid water essential for their well-being below freezing point; and b) they start breaking down above a few hundred degrees Celsius. This narrow range of temperatures makes them suitable for life on Earth only.

Besides carbon and water, life also favours left-handedness. Each amino acid, except glycine, exists in mirror-image left- and right-handed forms. But all the naturally occurring proteins in all organisms are made up of left-handed amino acids. Why this bias towards left-handedness? There are two explanations: a) amino acids that fall to Earth from space are more left-handed than right-handed because space radiation destroys more right-handed amino acids (mostly left-handed amino acids have been found in meteorites); and b) left-handed amino acids are more stable in water, and that's why proteins are made up of them (life, after all, started in water). The first explanation boosts the case of panspermia supporters; the second one deflates it.

Silicon-based life

Silicon also has some life-giving properties of carbon: it can form long chains to which other elements bind, and it is abundant in the universe (sand is silicon dioxide), though not as abundant as carbon. These peculiarities of silicon have led some scientists to suggest that there might be silicon-based life on other planets.

However, Arrhenius was totally against this speculation. He said in 1908: 'All organic beings in the whole universe should be related to one another and should consist of cells which are built up of carbon, hydrogen, oxygen and nitrogen. The imagined evidence of living beings in other worlds whose constitution carbon is replaced by silicon or titanium must be relegated to the realm of improbability.'

Carl Sagan, 'a carbon chauvinist' in his own words, said in 1976: 'Carbon compounds are not just more abundant but more stable. So, while life on other planets would probably not look like life on Earth because its internal biochemistry would be astoundingly different, I think it would be based on carbon.'

While scientists root for carbon, science fiction writers relish the idea of silicon-based life. Even the gifted H.G. Wells was 'startled by the visions of silicon-aluminium organisms … wandering through an atmosphere of gaseous sulphur'. More recently, in 2267, in fact, *Star Trek*'s Commander Spock has met silicon-based life, the Horta, on the planet Janus VI.

Carbon-based life is mainly made up of right-hand carbohydrates and left-hand amino acids. This handedness helps in biological processes. Silicon can't do it; it doesn't form man compounds having handedness.

Indecent haste

The debris of the Big Bang that marked the beginning of the universe began to fly away from the explosion point, is still flying and will keep on flying indefinitely. Time begins with the Big Bang, which happened 13.8 billion years ago.

Our planet was formed about 4.5 billion years ago from a ring of gas and dust around the young Sun. For nearly 700 million years, the young Earth was subjected to intense bombardment by gigantic asteroids; the debris leftover from the formation of the solar system. One such impact gouged a chunk out of Earth and formed the Moon. Life appeared about 3.8 billion years ago as soon as the cosmic bombardment had ended. Some biologists say life appeared 'fully formed, with almost indecent haste'.

The quick appearance of life on a young and inhospitable Earth suggests that life is easy to create, and therefore it could have easily arisen on other planets in the galaxy.

A 'wild and visionary' hypothesis

William Thomson (also known as Lord Kelvin) was the greatest physicist of his time. For 35 years, he was a professor at the University of Glasgow, but he was a failure as a lecturer and teacher. He was so preoccupied with his work that if any new idea came to his mind while lecturing, he would digress and forget all about the topic of his lecture. Yet he liked to illustrate his lectures with demonstrations. Once, to explain a point, he brought an old muzzle-loader rifle and shot it at a pendulum. On another occasion, he brought two eggs; apparently one boiled and one raw, to show the difference in how they behaved when spun. He said smiling when he began the demonstration: 'Both boiled, gentlemen.' He had quickly discovered that some mischievous student had decided to play a practical joke on him by secretly boiling both eggs.

No, he was not the archetypical absent-minded professor; he had an extraordinarily clear mind and a powerful personality. 'He was certainly inspiring to students,' according to one of his at the university. 'His enthusiasm was infectious.' He once said: 'Science is bound by the everlasting laws of honour to face fearlessly every problem that can be presented to it.' He fearlessly faced many problems of science.

In his presidential address at the 1871 annual meeting of the British Association for the Advancement of Science in Edinburgh, Thomson tackled the question of the origin of life on Earth.

From Aristotle to Newton, all old men of science accepted without any serious questions that life could arise instantaneously from living or non-living matter: maggots from decaying meat; caterpillars from leaves; frogs from slime. This view, called spontaneous generation, was first challenged in 1668 by Francesco Redi, an Italian physician and poet. Redi prepared eight flasks, each with a variety of meat in it: a dead snake, some fish and pieces of veal. He sealed four jars and left the other jars open to the air. After a few days, he found that only the open jars bred maggots. The meat in the sealed jars was just as putrid but without maggots. He now repeated the experiment, but this time covering four jars with gauze tops instead of sealing them. Air could enter the jar but not flies. Again, maggots appeared in the open jars only. From these experiments, Redi concluded that maggots were not formed by spontaneous generation but came from eggs laid by flies.

But even Redi's experiments could not shake people's belief in this age-old idea. The theories of Louis Pasteur and Charles Darwin finally laid to rest the idea of spontaneous generation. In 1862 Pasteur performed a series of experiments to answer 'the question of generation, so-called spontaneous': 'Can matter organise itself? In other words, can beings come into the world without parents, without ancestors? Here is the question to resolve.' He resolved the question by declaring that all life comes from another life. In 1859 Darwin published his theory of evolution of all present-day species from simpler forms of life through natural selection. Scientists did not readily accept Darwin's theory, but it renewed scientific debate about the origin of life.

In his address, Thomson said that science had brought a vast mass of inductive evidence against the hypothesis of spontaneous generation. 'Dead matter cannot become living without coming under the influence of matter previously alive. This seems to me assure a teaching of science as the law of gravitation.' He then addressed the question of the origin of life on Earth: 'Hence and because we all confidently believe that there are at present, and have been from time immemorial, many worlds of life besides our own, we must regard it as probable in the highest degree that there are countless seed-bearing meteoritic stones moving about through space. If at the present instant no life existed upon earth, one such stone falling upon it might, by what we blindly call natural causes, lead to its becoming covered with vegetation … The hypothesis that life originated on this earth through moss-grown fragments from the ruins of another world seems wild and visionary; all I maintain is that it is not unscientific.' His audience did not consider the idea scientific. They were sceptical of the notion of micro-organism-bearing meteorites flying in space and seeding different worlds.

Panspermia

However, at least one great man was impressed with Thomson's hypothesis. This admirer was Svante Arrhenius, the Swedish chemist who gave us the chemistry of ions, for which he won the 1903 Nobel Prize. Ironically, Thomson opposed the ionic theory when it was first proposed, saying that he could not understand anything which could not be translated into a mechanical model (for this reason, he had also rejected Maxwell's theory of electromagnetic waves). Somehow, the idea of life from outer space did fit into his mechanical and mathematical universe.

Arrhenius was a versatile genius who investigated many scientific ideas beyond chemistry. In 1896 he was the first to recognise that carbon dioxide acts as a thermal blanket around the globe, thus creating the greenhouse effect. In 1906, in a book translated into English as *World in the Making* (1908), he suggested that bacterial spores and other dormant micro-organisms escaped from another planet where life already existed, travelled through space, and finally landed on Earth and began to grow and develop. He called this process *panspermia* (Greek for 'all seeds').

Arrhenius did not explain how life originated on other planets. He just said that life is eternal. It has always been there, so the question of its origin does not arise. But he did support evolution: 'Life must always recommence from its very lowest type … and must pass through all the stages of evolution from the single-cell upward.'

The idea of panspermia fascinated scientists of the 19th century; however, it never became an accepted scientific idea. There are many modern versions of panspermia. Here's an interesting one.

Panspermia directed by ET

In 1971 Francis Crick and fellow molecular biologist Leslie Orgel attended the first-ever international scientific meeting on 'Communication with Extraterrestrial

Intelligence' in Soviet Armenia. At the conference banquet, their host, the astronomer Viktor Ambartsumian, called on each guest to propose a toast, after which they were expected to down a glass of vodka. 'I vaguely remember … my own toast to all extraterrestrial Armenians wherever they may be,' recalls Orgel. 'I also remember Francis Crick seeking relief from vodka, pouring a tumbler of what looked like water from a large jug, only to find it was more vodka. The young Russian student who came to help him told him not to worry, "the last Englishman to come to a party here had to be carried home".'

At that meeting, with or without the help of vodka, Crick and Orgel hit on the idea that perhaps life on Earth originated from micro-organisms sent here, on an unmanned spaceship, by an advanced civilisation elsewhere. 'We called our theory directed panspermia,' recalls Crick. 'Panspermia is the idea that micro-organisms drifted to the Earth through space and seeded all life on Earth. We used "directed" to imply that someone had deliberately sent the micro-organisms here in some way.'

Two years later, Crick and Orgel detailed their intriguing hypothesis in a paper in *Icarus*. They admitted that the chances of extraterrestrial micro-organisms reaching Earth either as spores driven by radiation pressure from another star or embedded in meteorites are extremely small. However, it is possible if someone decides to do it. Are you and I the result of an experiment conducted aeons ago by drunken ET nerds?

Two biological anomalies support the hypothesis. The first concerns the universality of the genetic code. There is only one genetic code for all forms of life. Why is it so? No satisfactory answer from molecular biologists has come up yet. The universality is due to a 'seed' sent by an extraterrestrial civilisation. As life originated from this 'seed', its genetic code replicated in all forms of life, resulting in a universal genetic code. That's the view of Crick and Orgel.

The second anomaly is the importance of molybdenum in biological systems. Many enzymes require the help of this trace element for their proper functioning. This would not have surprised molecular biologists if Earth were relatively rich in this metal, but it constitutes only 0.02 per cent of Earth's composition. Scientists expect the chemical makeup of living organisms to reflect the composition of the environment in which they evolved. 'If it could be shown that the elements represented in terrestrial living organisms correlate closely with those that are abundant in some class of – molybdenum stars, for example – we might look sympathetically at panspermia theories,' Crick and Orgel say.

They also discuss several questions likely to be raised against their hypothesis. Has there been enough time in the life of our galaxy for the sequential development of two advanced civilisations, one on Earth and the other on some planet beyond the solar system? Earth-like planets existed as much as 6.5 billion years before the solar system's formation, and 3.8 billion years elapsed between the appearance of life on Earth (wherever it came from) and the development of our technological society. Thus, the time available makes it possible that technological societies existed elsewhere in the galaxy even before the formation of Earth.

The other important question is about the possibility of safe transport of life over interplanetary distances. Research by other scientists has shown that life could probably be preserved for more than a million years if suitably protected and maintained close to absolute zero. This timeframe supports Crick and Orgel's calculations, which suggest that at a speed of 96,000 kilometres per hour, a spaceship packed with micro-organisms – blue-green algae, for example, has simple nutritional requirements – could infect most planets in the galaxy. However, they warn strongly that we should risk infecting other planets under no circumstances.

Even if we heed this warning, we could accidentally 'infect' other planets in the way that micro-organisms might have infected our own planet in the garbage left over by extraterrestrial visitors. This preposterous idea was suggested in 1960 by maverick American astronomer Thomas Gold. He imagined that interstellar visitors forgot to clean up after having a picnic on Earth. Do piles of garbage left in picnic spots worldwide hint that we have inherited this bad habit from our interstellar ancestors?

Crick revisited the notion of directed panspermia in his 1981 book *Life Itself: Its*

Origin and Nature. 'How *could* such stuff be considered seriously?' he asks. 'The whole idea stinks of UFOs or the Chariots of the Gods or other common form of silliness.' He replies: 'The kindest thing to say about directed panspermia, then, is to concede it is indeed a valid scientific theory, but that as a theory, it is premature.'

In the (primordial) soup

The eminent Scottish biochemist and geneticist, J. B. S. Haldane, would have dismissed the idea of an extraterrestrial seed as 'worthless nonsense'. For two reasons:

First, he once said (jokingly) that the normal process of acceptance of a scientific idea has four stages: i) this is worthless nonsense; ii) this is an interesting but perverse, point of view; iii) this is true, but quite unimportant; iv) I always said so.

Second, he was a firm believer in the idea that life originated on Earth. His ideas have helped shape the current picture of the origin of life.

Darwin avoided the question in his classic *The Origin of Species*, except in the final paragraph when he said that 'the Creator' originally breathed life 'into a few forms or into one', and then it evolved 'from so simple a beginning' into 'endless forms most beautiful and most wonderful'.

However, in a letter written in 1871 to his botanist friend Joseph Hooker, he suggested that life could have arisen, without supernatural intervention, 'in some warm little pond, with all sorts of ammonia and phosphoric salts, light, heat, electricity, etc. present', where it was possible 'that a protein compound was chemically formed'. This compound was 'ready to undergo still more complex changes; at the present day such matter would be instantly devoured or absorbed, which would not have been the case before living creatures were formed'. For the next century, Darwin's 'private hypothesis' dominated thinking on the subject.

In 1924 Aleksander Oparin, a 30-year-old Russian biochemist, presented the first viable theory of the origin of life by natural physical and chemical means here on Earth: in the early atmosphere, simple inorganic compounds combined to form complex organic compounds, which formed the first living cell.

During Stalin's reign, Oparin held a powerful position in the Soviet Academy of Sciences. Rumour has it that he used to sit down for dinner with a bottle of cognac on one side and a bottle of vodka on the other, both of which would be empty by the end of the meal. Oparin, who died in 1980, is admired not for his drinking prowess but for the 'primordial soup' he concocted.

He suggested that early in the Earth's history the atmosphere was rich in hydrogen. Simple inorganic hydrogen compounds such as water, methane and ammonia could form organic compounds. Gradually these organic compounds fell from the atmosphere to the ground, where the rain – which occurred when Earth cooled, and water vapour condensed – washed them into pools and ultimately into oceans. Over millions of years the organic molecules in this 'primordial soup' joined together into long chains of proteins and DNA molecules until a cell appeared which had the right kind of reactions and right kind of compounds to be considered an organism. This first cell could replicate itself and therefore it filled the bill for the first living organism.

In 1929 Haldane independently proposed the same general theory: 'The first living and half-living things were probably large molecules synthesised under the influence of the Sun's radiation, and only capable of reproduction in the particularly favourable medium in which they originated. Each presumably required a variety of highly specialised molecules before it could reproduce itself, and it depended on chance for a supply of them. This is the case today with most viruses ... which can grow only in presence of the complicated assortment of molecules found in a living cell.'

In 1953 Stanley Miller, a young graduate student working in the laboratory of Nobel Prize-winning chemist Harold Urey at the University of Chicago, provided the first experimental support to the 'primordial soup' theory. He subjected a mixture of methane, ammonia, water vapour and hydrogen to a series of electrical charges. He imagined this to be a rough duplication of conditions on the primitive Earth when the primordial soup was subjected to bolts of lightning.

After a week, the inorganic molecules had joined to form several amino acids, building blocks for proteins which make up cells. Urey was exultant: 'If God didn't do it this way, He missed a good bet.' (The Murchison meteorite described earlier was also shown to contain a number of the same amino acids that Miller identified.)

Miller's 'primordial soup' has been the staple diet of biology textbooks for decades but is now under serious threat. 'Although the taste of the soup seems quite agreeable now – at least with its broad experimental base – there are critics who go for other courses as far as the origin of life is concerned,' says Günter von Kiedrowski of the University of Ruhr.

Just one other course: Günter Wächtershäuser is a German patent attorney who also holds a doctorate in chemistry. His work has appeared in prestigious journals. He suggests that when we think of the origin of life we must not think of the DNA, the RNA or the cell. These things must have come later.

'Life didn't begin with a soup of chemicals, which cannot do anything, which are inactive,' he says. 'It's change that leads to life.' Natural chemical reactions provide that change. Life must have started in the simplest possible way, as a cycle of natural chemical reaction that repeated itself.

This cycle most likely started on the surface of pyrite; the shiny crystal known as fool's gold. It catalyses energy-production reactions known as the citric acid cycle that occurs in all organisms, including us. Wächtershäuser describes these reactions as a 'primitive metabolism': 'It's my theory that life began with a metabolism, and this metabolism invented everything else – genetic machinery and cells that enclose it.'

All this did not necessarily happen in Oparin's primitive oceans. Wächtershäuser claims that this process can happen anywhere: 'In vents under the ocean, in volcanoes on land, even deeper down, anywhere where gases come out from Earth. In my theory, the origin of life goes on today.'

Now we know that the primitive Earth was not as hospitable as Darwin's 'warm little pond' or Oparin's primitive oceans envisioned. Some scientists say that they were not favourable environments for the origin of life.

There are microorganisms that can grow at 110°C (230°F) in deep-sea volcanic vents 2.8 kilometres (1.7 miles) below the surface. These scientists believe that life got its start in geothermal pools. It is possible that some microorganisms might be living even now under the surface of Mars or the baking crust of Venus.

CHAPTER 6

A PARADOX TO BEHOLD

If they are there, why aren't they here?

The recent discoveries of thousands of exoplanets or extrasolar planets – planets beyond the solar system – have made the age-old question 'Are we alone?' highly tantalising.

In 1950 Enrico Fermi answered the question in his own inimitable way: by asking another question. Fermi, the most outstanding Italian scientist of modern times, was forced to flee Italy shortly after receiving the 1938 Nobel Prize for physics for his work on nuclear processes. He moved to the United States, where in 1942 he built the world's first nuclear reactor.

Fermi revelled in posing unexpected questions on aspects of the natural world and then figuring out their answers. According to American physicist Philip Morrison: 'Fermi was the first physicist to my knowledge who enjoyed doing physics out loud walking through the hall.' He describes an incident when they were walking through the wooden barrack-like structure of the theoretical physics building at Los Alamos Laboratory, and as they walked, the sounds of their footsteps reflected off the wooden surface and seemed to bounce throughout. Fermi asked Morrison: 'How far do you think our footsteps can be heard in this building?' Fermi then quickly started working out loudly what the yield of sound would be from the impulse, how far that would go, how the wood conduction and air passage would affect the sound, and by the end of the hall he had an answer. 'Sounded very reasonable,' says Morrison. 'And when I tried to recalculate, I got something like the same result – slowly and looking at the numbers repeatedly.'

Questions that can be answered quantitatively by rough approximations, inspired guesses and statistical estimates from little data are known as Fermi questions. Some classic Fermi questions: How many piano tuners are there in the city of Chicago? How many atoms could be reasonably claimed to belong to the jurisdiction of the United States? How far can a crow fly? These questions can be answered by making reasonable assumptions, not necessarily relying upon definite knowledge for an exact answer. (You can find many Fermi-type questions by Googling 'Fermi questions'. In the meantime, how many copies of this book would you need to fill your room?)

At a lunch in the summer of 1950 at Los Alamos, Fermi and fellow nuclear physicists Emil Konopinksi, Edward Teller and Herbert York were talking about space travel. The discussion was probably prompted by a cartoon by Alan Dunn in 20 May 1950 issue of the *New Yorker* magazine explaining why rubbish bins were disappearing from the streets of New York City. The cartoon showed 'little green men' with antennae carrying rubbish bins, marked DSNY, towards a flying saucer.

The discussion veered towards the possibility of many civilisations beyond Earth. Fermi surprised everyone by asking the provocative question: 'If they are there, why aren't they here?' This is Fermi's paradox.

What is the answer, then?

There are many explanations for Fermi's paradox:

Biological – Evolutionary biologist Ernst Mayr: The evolution path that leads to intelligent life is far more complex than we suppose – we are, if not the first, then among the first intelligent life forms to evolve in the galaxy.

Astronomical – Astronomer Carl Sagan: Daunting distances of interstellar space make space travel impossible … if we are alone in the universe, it sure seems like an awful waste of space.

Astrophysical – Astrophysicist Milan M. Ćirković argues that the universe is too hot right now for advanced, digital civilisations to make the most efficient use of their resources. They are sleeping and waiting for the universe to cool down, a process known as aestivating (like hibernation but sleeping until it's colder).

Bizarre – Astronomer John A. Ball: Zoo hypothesis which portrays Earth as a zoo of intelligent life in the galaxy – they're watching us from a distance.

Optimistic – Astronomer Frank Drake: 'They could show up here tomorrow'.

Humorous – Science writer Arthur C. Clarke: 'I'm sure the universe is full of intelligent life – it's just been too intelligent to come here'.

More explanations from known and unknown wits:

- We don't see them because they're all living near or even inside a black hole.
- We are doing a lackadaisical job in looking for them.
- They have visited, but not in recorded history.
- They have visited in recorded history (Erich von Däniken's *Chariots of the Gods*).
- They choose to exist in other dimensions.
- They are intelligent but lack the insatiable curiosity humans display and have no desire to communicate.
- They are omniscient, omnipotent beings and do not want to come here, as they fear we might start worshipping them like gods.
- They find travelling by spaceships boring; they prefer to meet by Zoom (using not the damn slow Wi-Fi of earthlings but their superior telepathy).
- They are couch potatoes with sophisticated remote controls (why leave your couch, let alone your planet?).
- This dispiriting view comes from Rod Liddle of UK's *The Spectator* magazine: 'They came, they had a look around and decided that it wasn't

for them after all – and that we're altogether too ghastly to warrant a waste of another nanosecond of space-time. And so, they have gone back wherever it was they came from, annoyed, depressed and possibly suicidal.'

A word to the wise

On a serious note, the question of life on other worlds has two aspects: a) is there any life form, even microscopic, on a planet other than our own; and b) is there intelligent life, able to communicate with us, on a planet in a star system other than our own? Both aspects have profound cultural, philosophical, scientific and religious implications.

And a warning from an extraordinary genius, Stephen Hawking: 'Meeting an advanced civilisation could be like Native Americans encountering Columbus. That didn't turn out so well for the Native Americans. Hawking and other critics of active search for intelligent aliens suggest to 'lie low', hoping that a technologically advanced civilisation on a mission to colonise new planets will simply not notice us.

In a letter to the editor, a *New Scientist* reader, John Bailey, writes that no one holds a democratic mandate to broadcast humanity's existence to potential hazards. 'Some may scoff at the vast distances, but I am sure the Incas and Aztecs would have felt that Madrid was a long way away,' he says candidly.

Dan Werthimer, a SETI researcher at University of California at Berkley, also think that active messaging to aliens is like inviting our own doom. 'It's like shouting in a forest before you know if there are tigers, lions and bears or other dangerous animals there,' he says.

CHAPTER 7

MAYBE THE ANSWER IS IN MATHS

Search for extraterrestrial intelligence (SETI)

American astronomer Frank Drake is a pioneering SETI researcher. In 1960 he became the first person to use a radio telescope to listen to ETs. In those early days of SETI, many scientists ridiculed the idea of extraterrestrial intelligent life, but to Drake, the idea of other intelligent civilisations beyond Earth was a distinct possibility. He even placed a sign on his office door at the National Radio Astronomy Observatory in Green Bank, West Virginia: 'Is there intelligent life on Earth?' 'People would stop to read the sign, and then slowly smile at the way it framed the extraterrestrial question. Often as not, a head would poke inside my doorway with a wry comment such as, "There's a little green man downstairs who says he's looking for you", or "Just checking to see if there's any intelligent life in this room",' he writes in *Is Anyone Out There?*, a book he co-authored with Dava Sobel.

Frank Drake's search for extraterrestrial intelligence continues and now carried out by about 100 scientists at the SETI Institute. Jill Tarter was its first scientist employee when the institute began operations in 1985. She is still devoted to the cause of searching for extraterrestrial intelligence.

In 2015, Israeli-Russian billionaire Yuri Milner, gave a tremendous boost to SETI research when he announced a 1—year program called Breakthrough Listen, funded by a $100 million donation. Breakthrough Listen is a scientific program searching for evidence of technological life in the Universe. It aims to survey one million nearby stars in our galaxy and 100 nearby galaxies.

Drake's equation – sheer speculation

In 1961 Drake invited a dozen scientists to the first-ever SETI conference (now remembered as the Green Bank conference). While he was preparing for the meeting, he came up with an equation consisting of astronomical, environmental, biological and cultural parameters to estimate the number (N) of advanced civilisations that exist now in our galaxy and are capable of communicating across interstellar distances:

$$N = R_* \times f_p \times n_e \times f_l \times f_i \times f_c \times L$$

where

R_* = rate of star formation in our galaxy

f_p = fraction of these stars with habitable planets

n_e = average number of planets suitable for life in each solar system

f_l = fraction of planets where life developed

f_i = fraction of planets where intelligent life developed

f_c = fraction of planets with technological civilisations

L = average lifetime of such communicating civilisations

This equation is now known as Drake's equation. 'It amazes me to this day to see it displayed prominently in most textbooks on astronomy, often in a big, important-looking box. I've seen it printed in *The New York Times*,' writes Drake in his book. Since then, the equation has become much more popular and ubiquitous. Even in the 21st century, it continues to influence scientific thinking about intelligent life on other worlds.

But can it provide an accurate estimate? Some scientists are not so sure and call it speculative in the extreme. Jill Tarter, a prominent SETI researcher (the character of Dr Ellie Arroway in Sagan's book and movie *Contact* was probably based on Tarter), says that it is 'really nothing more than a systematic way of quantifying our ignorance'. Nevertheless, if we assign values to the seven factors on the right-hand side of the equation, we can calculate N.

Here are some estimates made by participants at the Green Bank conference in 1961 and the current estimates:

R_* = Of all the seven factors, R_* is the only one supported by observational evidence beyond the solar system. Participants at the Green Bank conference estimated the value as 1. Current estimates vary from 1.5 to 3 stars per year.

f_p = The Green Bank group estimated this value to be about 0.5 (50 per cent), but ever-optimistic Sagan suggested a higher value of 1, which is also the most recently accepted value.

n_e = If we base our guess on the solar system, we can say that probably four worlds per solar system – Earth (absolutely), Mars (maybe in the past), Jupiter's moon Europa and Saturn's moon Enceladus (remote possibility) – are capable of sustaining life. Using our solar system as a guide, the Green Bank group estimated between 1 and 5 planets. The current estimate is roughly 0.4.

f_l = Some scientists argue that this value is 1 because life is virtually inevitable on any habitable world; others give it a pessimistic value of one in a million, as they believe that the chances of life appearing on a planet or moon are extremely small. The Green Bank group also claimed that, given time, life will appear in a suitable environment and assigned it a value of 1, which is also the currently accepted value.

f_i = The Green Bank group concluded that wherever there was life in the universe it ,would show signs of intelligence and gave f_i a value of 1. The current estimates vary from a pessimistic 0.0002 (based on the very long period it took intelligent life to evolve on Earth) to an optimistic 1 (once primitive life appears, it will surely evolve into an intelligent form).

f_c = The Green Bank group, however, arrived at an estimate of 0.2. This value can be taken as 1 on the simple reasoning that once you have intelligent life like ours, it's likely to develop capabilities to communicate at interstellar distances.

L = This is the most uncertain value in Drake's equation. We don't have any basis for a reliable estimate; even the only known technological civilisation has been communicating with radio waves for only about 100 years. Will our civilisation last a million years, or will we destroy ourselves in the near future? Therefore, the value of L can range from an optimistic 1,000 to 100 million years (the Green Bank group's estimates) to a cautious 300 to 10,000 years (current estimates).

Now, by multiplying the seven terms we can arrive at a value for N and these values vary from 1 to many millions. This shows only what we already know: either we are alone or there are a large number of worlds with intelligent life that can communicate with us. Drake's equation is simply a mathematical way of saying 'we don't know'.

The following photographs show Dr Frank Drake explaining his famous equation. (Photo: Dr Seth Shostak, SETI Institute, California)

86

Quantifying our ignorance with other equations

Astronomers Fred Hoyle and Chandra Wickramasinghe say that the number of technologies at any time is given by the formula:

number of technologies = technological lifespan in years ÷ (life of a typical star ÷ number of habitable planets

With 2 billion habitable planets and an average life of a typical star of 10 billion years, this gives: number of technologies = technological lifespan in years ÷ 5

If 300 years is taken as the typical technological lifespan, we obtain only 60 technologies throughout the galaxy. If we take a much more optimistic estimate of 300,000 years, which is about the length of time *Homo sapiens* has been in existence: 'In this case the number of technologies in the galaxy turns out to be 60,000, with a grand total of 60 million for the entire visible universe,' they say.

Peter Ward and Donald Brownlee, authors of *Why Complex Life Is Uncommon in the Universe*, pour their *Rare Earth* cold water on this optimistic estimate by their intimidating equation:

$$N = N^* \times f_p \times f_{pm} \times ne \times ng \times f_i \times f_c \times f_l$$

where

N^* = stars in the Milky Way galaxy

f_p = fraction of stars with planets

f_{pm} = fraction of metal-rich planets

ne = planets in a star's habitable zone

ng = stars in a galactic habitable zone

f_i = fraction of habitable planets where life does arise

f_c = fraction of planets with life where complex animals arise

f_l = percentage of a lifetime of a planet that is marked by the presence of complex animals

They say that they have left out some of the more exotic aspects of Earth's history, such as plate tectonics and a large moon, from this equation. The point that Ward and Brownlee are making is that in any equation when a term approaches zero, so does the final result. If a planet does not have a right moon or its surface lacks conditions that can support water, then $N = 1$.

American astrophysicist Michael H. Hart has a better way to estimate N, without any equation: Assume some value for N and inquire about its effect on observables. In particular, he asks, if N is very large, then 'why aren't they here?' If there are a million civilisations in the Milky Way, he says, then at least one of them would have colonised our solar system by now, since the time to colonise the entire galaxy – with nuclear rockets, for example – is less than the age of the galaxy.

Seager's equation – Drake's speculation updated

In 2017, Sara Seager, the American astronomer whose biosignature research was described in chapter 3, updated Drake's 1961 equation based on what we now know about exoplanets. Drake's equation assumes that extraterrestrials are intelligent and use radio technology. Seager assumes that life of any type present in sufficient abundance can change the chemical composition of its planet's atmosphere.

Drake's equation has seven factors which multiplied together give the number of intelligent alien civilisations we could hope to find. In Seager's equation, which has six factors, the number of planets with detectable signs of life:

$$N = N_* \times F_Q \times F_{HZ} \times F_O \times F_L \times F_S$$

where

$N^* =$ the number of stars that will be observed

$F_Q =$ the fraction of those stars with habitable planets

$F_{HZ} =$ the fraction of stars with rocky planets in the habitable zone

$F_O =$ the fraction of those planets that can be observed

F_L = the fraction of planets that have life

F_S = the fraction of planets with life that produce a detectable biosignature gases

Sun-like stars are not the only stars that have habitable planets. Seager's equation focuses on red dwarf stars, which are cooler and smaller than the sun. The nearest Earth-size planet orbit a red dwarf 6.5 light-years away.

Seager calls her equation Biosignature Drake's Equation, and just like with Drake's equation, some of the factors in Seager's equation are speculative. She has assigned the following values to six factors: $N* = 81$, $F_Q = 0.8$, $F_{HZ} = 0.1$, $F_O = 0.67$, $F_L = 0.5$, $F_S = 0.5$. Multiplied together, N, the number of planets with detectable signs of life, equals about 1. It looks like we will never meet ET.

More maths but no equations

Adam Frank, an American astrophysicist, believes that the revolution in planetary knowledge and the ever-increasing number of exoplanets being discovered demands a fresh look at Drake's equation. The equation was not a statement of a universal law like is not like Einstein's $E = mc^2$, he says, it was a mechanism for fostering organised discussion.

'Three of the seven terms in Drake's equation are now known,' he writes in an article, 'Yes, There Have Been Aliens', in *The New York Times*. 'We know the number of stars born each year. We know that the percentage of stars hosting planets is about 100. And we also know that about 20 to 25 percent of those planets are in the right place for life to form. This puts us in a position, for the first time, to say something definitive about extraterrestrial civilizations – if we ask the right question.'

Without going into details of his maths wizardry, his answer is: 'While we do not know if any advanced extraterrestrial civilizations currently exist in our galaxy, we now have enough information to conclude that they almost certainly existed at some point in cosmic history.'

The question which remained to be answered: Dead or alive, did they use antimatter to fire up their UFOs to spy on us? Star Trek fans know that Starship Enterprise is powered by antimatter. Antimatter is no longer stuff of science fiction; it does exist.

Let's roll a dice to find a cosmic neighbour

If you find the idea of being alone in the universe gloomy, then turn to American statistician and author Amir D. Aczel for help. He has written a whole book on Drake's equation, *Probability 1: Why There Must Be Intelligent Life in the Universe*, to show that there is intelligent life on at least one other planet.

First, Probability 101. Chevalier de Méré, a 17th-century high-living French nobility and gambler, liked to bet even money that a 6 would come up in four rolls of a die. But his luck changed when he started betting even money that a 6 would come up at least once in 24 rolls with two dice. He asked his friend, Blaise Pascal, one of the most brilliant mathematicians of his day, why he was having bad luck in his new game. Pascal wrote to fellow mathematician Pierre de Fermat about this problem, and their correspondence on the matter led to the birth of probability theory.

Chance is something that happens unpredictably. Probability is the mathematical concept that deals with the chances of an event. We can find the probability of an event by dividing the number of ways the event can happen by the total number of possible outcomes. A dice has six faces, numbered 1, 2, 3, 4, 5 and 6. The probability that any one of these numbers comes up is $1/6$.

What would be the probability of getting either a 3 or a 5? Because 3 and 5 cannot occur together, such an event is called a mutually exclusive event. In such events, the probability is calculated by adding individual probabilities. The probability of getting either a 3 or a 5 is $1/6 + 1/6 = 1/3$.

When two dice are rolled separately, the second dice does not consider what the first dice has done to decide what it will do. Such an event is called an independent event. In independent events, the probability is calculated by multiplying independent probabilities. When two dice are rolled separately, the chance of getting a double 6 is $1/6 \times 1/6 = 1/36$. The probability of the complement of an event equals 1 minus the probability of the event. In other words, the probability of getting no double is $1 - 1/36 = 35/36$. (The complement of an event is its opposite; for example, if 'the coin shows heads' is the event in tossing a coin, then the complement is 'the coin shows tails'.)

After this one-minute lesson in probability, we're ready to look at de Méré's problem.

Single dice

Probability of a $6 = 1/6$

Probability of a number other than $6 = 5/6$

Probability of no 6 in four rolls $= 5/6 \times 5/6 \times 5/6 \times 5/6 = 0.48$

Probability of at least one 6 in four rolls $= 1 - 0.48 = 0.52$ or 52 per cent

De Méré's chances of winning his bet were 52 per cent. The odds, as punters say, were in his favour.

Two dice

Probability of a double $6 = 1/36$

Probability of no double 6s $= 35/36$

Probability of no double 6s in 24 rolls $= (35/36) \times (35/36) \times \dots$ multiply 24 times $= 0.51$

Therefore, probability of at least one double 6 in 24 rolls $= 1 - 0.51 = 0.49$ or 49 per cent

De Méré's chances of winning his bet were 49 per cent. The odds were against him. What are our odds of winning a cosmic neighbour?

In his book, Aczel takes the fraction of stars with planets, $f_p = 1/2$. When he wrote the book, only nine exoplanets had been discovered, and only one was in the habitable zone. He says that the fact that out of nine exoplanets, one is in the habitable zone, and this is confirmed in our solar system. Similarly, Earth is in the habitable zone, the other eight planets possibly not, so he says we will use $1/9$ for this parameter. He considers DNA 'an extremely complex molecule with a minimal chance of occurring on its own and that life is precarious because the universe is a dangerous place'. By this reasoning, he assumes the probability of life occurring on any planet already within a star's habitable zone to be extremely remote. He assigns it an arbitrary value of one in a trillion. By multiplying $1/2$, $1/9$ and $1/\text{trillion}$, he arrives at a figure of 0.000,000,000,000,05, which is the probability of life around anyone given star. He assumes that there are 300 billion stars in our galaxy and 100 billion galaxies in the universe. He then puts all these values in the rule for combining the probabilities of independent events:

The probability of life on at least one other planet outside Earth with life on it = 1 – (0.999,999,999,999,95) × (0.999,999,999,999,95) × ... multiply 30,000 billion billion times = a number very close to 1.

'While we used the best scientific estimates, even lower values still lead to the same answer, a number close to 1,' Aczel says. 'The probability is a virtual certainty.' We are not alone.

Terrestrial principles may be alien to the aliens.

Richard Dawkins, a world-renowned evolutionary biologist, now better known for his book, *The God Delusion*, claims that 'the Darwinian law may be as universal as the great laws of physics', and complex structures found anywhere in the universe are of living origin. Elling Ulvestad, a medical researcher at Haukeland University Hospital in Norway, has applied this 'theoretical framework' to arrive at an intriguing hypothesis: extraterrestrial radio signals are not necessarily generated by aliens.

His argument runs something like this: We can envision extraterrestrial radio signals as artefacts generated by intelligent life on other planets (hypothesis H_1). Although the theory of evolution is opposed to the design argument for explaining biological diversity, it's not opposed to using it for explaining artefacts. As such, the design argument is valid for scientific inference. Therefore, radio signals could be looked upon as valuable signs of meaning in the universe. However, when regarded as a scientific hypothesis, we have little reason to believe H_1 rather than the opposite hypothesis – that the signals are not generated by an alien civilisation (H_2). Since there is no valid data to support either hypothesis, both hypotheses have an equal chance of being right: probability (signal/H_1) = probability (signal/H_2).

Why do we then believe that the likelihood of H_1 is greater than H_2 – signals coming from intelligent life? Our Earth-centric background knowledge inclines us to believe it, explains Ulvestad.

He says that our ever-increasing technological sophistication and an almost obsessive wish to learn whether we are alone in the universe may well turn astrobiology into the major scientific enterprise of this millennium. The laws of physics may be universal, but astrobiologists have no valid pre-understanding of their subject matter. 'Since there is no assurance that life beyond Earth will be Earthlike, terrestrial biological principles and data do not necessarily provide valid information when inferring life in other worlds. Paradoxically, extrapolation of terrestrial principles to other worlds may preclude any chance of finding new principles of life.'

Ulvestad solves this paradox by suggesting that unprejudiced information about alien life forms can be gained only by applying biosemiotics (*bio*, life; *semion*, sign), the study of signs, of communication, and of information in organisms. He says that semiosis, any form of activity or process that involves signs, is 'a universal characteristic of living organisms because without semiosis, there can be no recognition … Semiotic communication involves the sign, the object that the sign refers to, and the interpretant. For something to be a sign, it must be understood as such – a sign is a sign only in context. Signs must be interpreted about each other in a context. Otherwise they may not even be acknowledged as signs.'

SETI researchers use radio signals, he continues his argument, because of the practical convenience of radio waves. Their Earth-centric presumption expects that aliens are equipped with receivers among their sense organs that respond to the same auditory signals that humans do. Furthermore, since the reception of any message is dependent on prior knowledge of the possibilities, he says, it is expected that the aliens have an evolutionary history similar to ours. 'I find both assumptions incomprehensible, and consequently, find the utilization of radio waves as a means for contact with extraterrestrial intelligence dubious also from an evolutionary angle.'

Aliens have evolved into machines

If we ever meet our sole neighbours on a world beyond Earth populated by Aczel's statistical wizardry, they're likely to be machines, not creatures made of carbon and water. That's if we believe Steven J. Dick, an American astronomer and historian of science.

Olaf Stapledon, the British author and philosopher who died in 1950, is remembered primarily for his fictitious 'histories of the future'. His science-fiction novel *Last and First Men* (1930) traces the history of humanity from the First Man (that's us) through the Eighteenth Man or the Last Man (living on Neptune 2 billion years hence).

Dick says that if biology and culture exist beyond Earth, they will evolve, and to understand this evolution we must think not only in astronomical timescales but in what he calls Stapledonian timescales, which consider the evolution of biology and culture. Applying Stapledonian thinking to biological and cultural evolution will make extraterrestrial intelligence different from ours.

He believes that we inhabit a post-biological universe, a universe in which the majority of life has evolved beyond flesh and blood intelligence. His post-biological universe is based on three premises:

1. The maximum age (A) of extraterrestrial intelligence is several billion years.

2. *L*, the term in Drake's equation, is greater than 100 years and probably much larger.

3. In the long term, cultural evolution overtakes biological evolution.

If we assume that we can survive nuclear world wars, mass extinctions and other natural catastrophes such as asteroid impacts, '*L* could conceivably approach *A*, which is billions of years'. Dick admits that we cannot know with certainty 'how long does *L* have to be before we reach a post-biological outcome?' By extrapolating current trends in artificial intelligence, biotechnology and nanotechnology, he estimates that we will undoubtedly make the transition to a post-biological universe if *L* is greater than 1,000 years. He accepts that this result is based on applying the insights of many others to the entire universe, using Stapledonian thinking.

What would this post-biological universe be like? Like the 1973 movie, *Westworld*, filled with gunslinging Yul Brynner robots? We hope not. Dick lists three characteristics of post-biological 'humans'. First, they will be immortal; they will have the capability to repair and update. Second, their capacity for great good or evil. Third, as nothing in the universe remains static, the post-biologicals will also evolve into creatures that 'might have all the characteristics we ascribe to God: omniscient, omnipotent and perhaps the capability of communication through messenger probes'.

Welcome to a Futureworld of 'self-improving thinking machines'. Where are all those cute ETs? Steven J. Dick, please step aside. Let Steven Spielberg write the scenario for the post-biological universe. Meanwhile, we earthlings shall wait for a whisper from space.

CHAPTER 8

RIDDLES AND MYSTERIES

Since they're not here, they do not exist

No, we will never get a chance to talk to an alien, if we accept American astrophysicist Michael H. Hart's bold argument for the absence of extraterrestrials on Earth: 'If there were intelligent beings elsewhere in our galaxy, then they would eventually have achieved space travel, and would have explored and colonised the galaxy, as we have explored and colonised Earth. However, they are not here; therefore, they do not exist.' Before we say QED, let's ask the good scientist to demonstrate his proof.

In an imaginative paper first published in 1975, Hart starts his case by labelling the statement 'There are no intelligent beings from outer space on Earth now' as Fact A. He says that the argument presented in the previous paragraph is basically correct but clearly incomplete. He puts all other explanations of Fact A into four groups:

1. *Physical explanations*: He dismisses the claim that aliens have never arrived on Earth because some biological or technological difficulty makes space travel impossible. For example, travelling at 10 per cent of the speed of light (about 1 billion kilometres per hour), a one-way trip to Sirius, one of the nearest stars, would take 88 years. He suggests that such large travel times may not be a problem to aliens if they have a lifespan of thousands of years. For aliens with a lifespan of 3,000 years, a voyage of 200 years is no different from the long sea voyages of our early explorers. Even if aliens do not have a long lifespan, it's not impossible to complete an interstellar journey in more than one generation

'if the spaceship is large and comfortable, and social structure and arrangements are planned carefully'.

2. *Sociological explanations*: He puts most proposed explanations of Fact A into this category. Three typical examples are the *contemplation hypothesis* (mature civilisations are primarily concerned with artistic and spiritual pursuits and are not interested in visiting or colonising faraway planets); the *self-destruction hypothesis* (all advanced civilisations blow themselves up not long after they discover nuclear weapons); and the *zoo hypothesis* (we live in a zoo of an advanced civilisation). 'No such hypothesis is sufficient to explain Fact A unless we can show that it will apply to every race in the galaxy, and every time,' he stresses.

3. *Temporal explanations*: They are not here simply because none have yet had the time to reach us. To judge the plausibility of this explanation, Hart assumes that we are indeed the first species in our galaxy to achieve interstellar travel. If we establish colonies on the 100 nearest stars, each of these colonies would establish their own colonies, and so on. He calculates that the time to traverse most of our galaxy would be about 650,000 years. 'We see that if there were other civilizations in our galaxy, they would have had ample time to reach us,' he says.

4. *Perhaps they have come*: Hart discusses three versions of this hypothesis. The most common version is that visitors from space arrived here in the fairly recent past (within, say, the last 5,000 years) but did not settle here permanently. The weak spot of this version, according to Hart, is that it fails to explain why Earth was not visited earlier. The second version of the hypothesis is that they visited us a long time ago, say 50 million years ago. The third version, the UFO hypothesis, is that aliens have not only arrived on Earth but are still here. Hart rejects this version: 'This version is not really an explanation of Fact A, but rather a denial of it.'

If his proof of Fact A is correct, Hart concludes, in the long run, our descendants will probably occupy most of the habitable planets in the galaxy. These descendants might eventually encounter a few alien civilisations which never took any interest in interstellar travel, 'but their number should be small, and could well be zero'.

It's sad, but we're totally alone

American cosmologist Frank Tippler also believes that if extraterrestrials exist, they should already be here. 'Since they are obviously not, they do not exist,' he says. He does not deny the possibility that primitive life is widespread in the universe but believes that the development of intelligence is vastly improbable. It has happened only once since the Big Bang. We are the first intelligence to evolve in the whole universe. 'We are totally alone,' he says.

Tippler's argument assumes that interstellar travel by self-replicating space probes is possible. If intelligent life were common, its emergence should have had a head start on planets around stars billions of years older than our Sun. At least one alien civilisation would have developed self-reproducing space probes and launched them into space. Travelling at 90 per cent of the speed of light – a technology not beyond the capabilities of an advanced civilisation, according to Tippler – it would take five years to reach the nearest star. If it takes 100 years to make a copy of itself, then the average speed at which all the probes would spread would be about $^{1}/_{25}$ of the speed of light. At this speed, Tippler estimates, the probes would spread throughout the galaxy within 10 million years. But we have no evidence of these probes on Earth. Their absence shows the absence of aliens. That's the logic of Tippler.

The idea of self-replicating machines was developed in the 1950s by Hungarian American mathematician and physicist John von Neumann. Self-replication would be possible if the hypothetical machine – von Neumann called it a 'universal constructor' – were provided with its description and a means of copying and transmitting this description to the newly constructed machine. Such a machine has two components which are wholly distinct from one another – the machine and its description. The description serves two purposes: first, as a program that is executed during the construction of the new machine; second, as passive data that is duplicated and given to the new machine. Put simply, the machine has two parts: a computer and a constructor. The computer directs the constructor. The constructor, in turn, manufactures both another constructor and another computer. The original computer then copies its program to the new computer, which starts executing the program.

Von Neumann never applied his idea to space probes, but many scientists including Tippler have done so since then. Tippler envisages his von Neumann probe as having human-level intelligence. It will be capable of using raw materials and energy resources available in distant star systems to replicate. On reaching a solar system, the probe will build several copies of itself. These copies will search for other solar systems, where the process will be repeated. 'Eventually, all the stars in the Galaxy would be reached by some descendant of the single original probe,' says Tippler. 'The Galaxy would be explored for the price of one von Neumann machine!'

His concept of a von Neumann probe goes a step further than simply making copies of itself to send to other stars. The probe would also be able to exploit mineral and energy resources of the star system to make technological artefacts, but he has no idea of the nature of these artefacts. Marcus Chown, a British science writer, agrees that this is not unreasonable: 'After all, the Romans would have had no idea that future civilisations would turn sand into computers, or bauxite into aeroplanes.'

These von Neumann probes could also be programmed to colonise space with replicas of humans. 'All the information needed to synthesise a human being is coded in the DNA of a fertilised human egg cell,' he says. 'This information could, in principle, be stored in the memory of a von Neumann machine, which could be instructed to synthesise an egg and place the "fertilised cell" in an artificial womb.'

Thus, our descendants one day will spread and colonise the entire universe. That's the dream of Tippler.

They could show up here tomorrow

Why are we fascinated by life beyond Earth? Aristotle answered this question more than two millennia ago when he said: 'All men by nature desire knowledge.' In our millennium, physicist Paul Davies has said: 'For in a sense, the search for extraterrestrial life is really a search for ourselves – who we are and what our place is in the grand sweep of the cosmos.'

Our place in the grand sweep of the cosmos will never be the same, even if we discover some fungus on another planet. It will show that probably life is widespread throughout the universe. The ages-old hubris that our planet is special will disappear. Not completely. Some will claim that humans are still the only intelligent life.

Though science has made great strides in finding primitive life, or at least likely abodes of such life, our searches are still limited by the terrestrial concept of life and intelligence. We like to look for things like ourselves.

SETI researchers keep on reminding us that 'absence of evidence is not evidence of absence'. Perhaps that's why Frank Drake, the pioneer researcher, is optimistic that 'they could show up here tomorrow'.

ETs will always have someone to talk to when they visit Earth

American astrophysicist J. Richard Gott is an expert on black holes, regions of space where gravity is so strong that nothing, not even light, can escape. He also loves using the 'Copernicus principle' to forecast the future duration of events as mundane as a couple's current relationship or as momentous as the longevity of the human race.

Copernicus rejected the prevalent belief that Earth stood still at the centre of the universe. The Copernicus principle says that one's location is unlikely to be special, or to put it bluntly, wherever or whenever we are, it's nothing special.

The maths of the Copernicus principle, according to Gott, works as follows: 'If there's nothing special about your observation of something, then there's a 95 per cent chance that you are seeing it during the middle 95 per cent of its observable lifetime, rather than during the first or last 2.5 per cent. At one extreme, the future is only $1/39$ as long as the past. At the other, it is 39 times as long. With 95 per cent certainty, this fixes the future longevity of whatever you observe as being between $1/39$ and 39 times as long as its past.'

In 1969 Gott visited the Berlin Wall, which was then eight years old. He applied the above formula (with 50 per cent likelihood; in a later version, he changed this value to a greater certainty of 95 per cent) and predicted that it would last more than $2^2/3$ years but less than 24. The Wall came down 20 years later in 1989.

Before you say 'wow', let's apply Gott's maths to *Homo sapiens*, who appeared about 200,000 years ago. It means that we'll last at least another 5,100 years but less than 7.8 million years. Turn the music on – there's no need for gloom and doom, yet. And life on Earth, which appeared about 3.8 billion years ago, will survive between 97 million and 148 billion years. We may have gone, but ETs will always have someone to talk to when they visit Earth.

Right now, we have only one example of life. One example could be a fluke, says John A. Ball – 'with two examples we can do statistics'. Even if there were one other world with life, the right statistics would be: one plus one does not mean two, but many worlds with life.

The statistics of Gott's theory is stark when we apply his theory to his own theory. It was first published in 1993; therefore, it should have future life of between two months and 156 years before it's sucked into a black hole.

The silent song of Oumuamua

Oumuamua is neither a riddle nor a mystery, simply astronomers' hype of astronomical proportions about an object in space. No, it was not a spaceship packed with cuddly aliens who would upon reaching Earth burst into a welcome song to earthlings: 'We'd like to get to know you/We know we'd get to like you/We hope you like us'. Yes, it was definitely the first space rock of its kind seen by humans, and it was clearly coming from outside the solar system.

In 2017 astronomers at the University of Hawaii saw a strange skinny object about 400 metres (1300 feet) long and only about 40 metres (130 feet) wide. It appeared either flat or cigar-shaped, dark and red. It was not a comet as it showed no tail. It could not be an asteroid as most asteroids are pock-marked potato-shaped rocks.

The rock was also strange in the sense unlike asteroids and comets which swing around the Sun in elliptical orbits with an average speed of about 64,000 kilometres (40,000 miles) per hour, it was hurtling in space up to speeds of 315,400 kilometres (196,000 mile) per hour.

The astronomer concluded it was an interstellar object, an object from a far-off solar system. They named it 'Oumuamua' (pronounced 'oh-mooah-mooah'), Hawaiian for 'messenger from afar'. However, this messenger from outer space is not sending out any signals.

In a paper published in the prestigious journal *Nature*, 14 astronomers concluded that Oumuamua was natural, and they found no compelling evidence to favour an alien explanation. However, some astronomers still speculate that Oumuamua was an artificial object, that is, a spaceship. One of them is astrophysicist Avi Loeb. He presents his case in an Opinion piece in *Scientific American*: 'Oumuamua was an artificial object on a targeted mission towards the sun, aimed to collect data from the habitable region near Earth. One might even wonder whether Oumuamua might have been retrieving data from probes that were already sprinkled on Earth at an earlier time. In such a case, Oumuamua's thin, flat shape could have been that of a receiver.'

It is intriguing that a renowned astronomer believes that Oumuamua was our first glance of a 'messenger' spaceship from an advanced alien intelligence. Even NASA admitted that it was definitely an unusual object. Unusual doesn't mean an object crafted by an alien civilisation.

Loeb knocks the phrase 'extraordinary claims require extraordinary evidence', popularised by Sagan, on its head when he says, 'extraordinary conservatism keeps us extraordinarily ignorant'. ET, please take note that, openminded as we earthlings are, we have duly noted your visit in our annals of encounters of alien kinds.

We can now say goodbye to Oumuamua as it is fast zooming away from us and now too far away to any more observations – and speculations.

CHAPTER 9

HOW WOULD INQUISITIVE ALIENS FIND US?

Biosignatures

How would we know if extraterrestrial life exists now or existed in the past out there in the universe? Walking in a remote area, when we see footprints, feathers, excrement or seeds flying in the air, we immediately know that we have found life. These are biosignatures of life. A biosignature is any characteristic, element, molecule, substance or feature that can be used as evidence of past and present life. It can also be a fossil, etching or layering in rocks or presence of water on land or in an atmosphere.

Another important biosignature is the ratio of isotopes of chemical elements. An isotope of an element has different numbers of neutrons in the nucleus. All the isotopes of one particular atom have the same number of protons but have different numbers of neutrons. For example, chlorine gas consists of two isotopes, chlorine-35 and chlorine-37. The nucleus of each isotope has 17 protons but a different number of neutrons: 18 in one and 20 in the other. More energy is used for metabolism – the chemical processes that occur within an organism – when the organism uses molecules that have lighter isotopes. Thus, by finding the ratio of the lighter isotopes to the heavier isotopes, scientists can conclude the existence of life.

Molecules that exist as gases can also be possible biosignature. For smart alien astrobiologists looking for life, Earth's atmosphere 21% oxygen concentration is a sure sign that there is life on Earth. The abundance of methane, methyl chloride, nitrous oxide and ozone in our atmosphere makes them easily detected from space.

Sara Seager, an astronomer at the Massachusetts Institute of Technology, has compiled a whopping list of 14,000 potential biosignatures in the atmospheres of alien planets playing host to life. Williams Bains, a biochemist at the University of Cambridge, goes even further when he says that conceptually there is no limit to what sort of gas life can produce.

Seager puts all possible biosignatures in three categories: a) gases such as methane released when organisms use chemical reactions to produce energy; b) gases like carbon dioxide and oxygen used in chemical reactions to build biomass; and c) everything else such as scents, perfumes and poisons.

Gases in the third category are ideal biosignatures for hunting ETs as terrestrial life makes smaller amounts of them and there are fewer chances of making them by geological processes. This category includes pungent gases such as: a) methyl chloride, a toxin produced in significant quantities by algae or marine bacteria; b) methyl bromide, produced by seaweed; c) dimethyl sulphide, which smells like cooked cabbage or Brussels sprouts; d) isoprene, a gas made by trees which smells like petrol; and e) methanethiol, comes from bad breath and flatulence.

Any unusual concentrations of these gases on any extraterrestrial planet will be a sure sign of cabbage-loving or foul-smelling ET.

Lisa Kaltenegger, a professor of astronomy at Cornell University, prefers looking for Freon, an artificial odourless gas used in refrigeration. She suggests that astronomers look for Freon in the atmosphere of distant planets. The abundance of Freon in a planet's atmosphere would suggests that the inhabitant love their drinks and food ice cold. There is only one problem with this idea: the overuse of Freon could destroy the planet's atmosphere. Certainly, a clever way to find dead or drunk aliens.

Well, the idea of looking for dead aliens has its supporters. Some astronomers suggest looking for 'extinction signatures' such as rotting corpses because of a nuclear war or a deadly pandemic.

Astronomical biosignatures

Extraterrestrials would probably find us by applying the techniques employed by our own planet-hunters. Once they found Earth, they would look for astronomical biosignatures (signs of life visible from space) to determine if there was any life on their newly discovered planet.

If you used a Bunsen burner in your school days, you would find it easier to understand the science behind astronomical biosignatures. The Bunsen burner was developed in 1855 by Robert Bunsen, a great German teacher and experimenter. Bunsen was a friend of Gustav Kirchhoff, who was a professor of physics at Heidelberg. Bunsen and Kirchhoff together developed the first spectroscope, a device used to produce and observe a spectrum. In 1860, Kirchhoff made the momentous discovery that, when heated to incandescence, each element produced its own characteristic lines in the spectrum. This means that each element emits light of a certain wavelength. In a leap of intuition, he even went further: anything any atom emits it must also absorb. Sodium's spectrum has two yellow lines (wavelengths about 588 and 589 nanometres). The Sun's spectrum has many dark lines, some of which correspond to these wavelengths. This means that sodium is present in the Sun. Scientists now had a tool they could use to find the presence of elements in stars. According to Isaac Asimov, Kirchhoff's banker was not impressed with his ability to find elements in the Sun. 'Of what use is gold in the Sun if I cannot bring it down to Earth?' he asked. When Kirchhoff was awarded a gold medal for his work, he handed it to his banker and said: 'Here's gold from the Sun.'

If inquisitive aliens have scientists as smart as Bunsen and Kirchhoff, they will know how to mine gold from Earth by passing light through a spectroscope. From the resulting spectrum they will find out that the Earth's atmosphere has oxygen, water, carbon dioxide and ozone, suggesting the presence of life. If they find out that the atmosphere is oxygen-rich, they will know that photosynthetic life constantly replenishes the supply.

If they detect Freon and other chlorofluorocarbons (from air-conditioning and refrigeration systems), methane (from sewage, rice cultivation, leaks from natural gas distribution and burps of domestic livestock) and smog (from fossil fuels) in the atmosphere, they will also know that their newly discovered planet is home to a high-tech society – intelligent enough to destroy itself by global warming.

Cosmic technosignatures

The story of the discovery of Martian canals by the American astronomer Percival Lowell described in chapter 2, though erroneous was the first attempt to look for signs of technology that might point to life in the universe. These signs of technology are known as technosignatures.

Artificial radio signals, as distinct from natural sources of radio waves, is an important technosignature. In 1960, using the Green Bank Telescope in West Virginia, Frank Drake made the first attempt to search for artificial radio signals from around two stars. In a similar way, inquisitive aliens might search for artificial radio signals.

If aliens have large radio telescopes, something like the Arecibo, they will receive radio waves from Earth, but they might find it difficult to distinguish these waves from those emitted by larger planets and the Sun. The signals from early television shows of the 1930s have so far travelled nearly 100 light-years, too short a distance to reach another civilisation. Besides, these signals can be detected only if alien eavesdroppers tune to the right frequencies.

Light pollution on the night side of a planet is also a possible sign of a technological civilisation. Avi Loeb, a renowned American astrophysicist, believes that there is a detectable difference between a planet shining with reflected light and a planet glowing with its own artificial illumination 'The search for city lights on habitable planets may sound speculative, but it is worth pursuing as a potential technosignature with planned instruments,' he says

Another technosignature is reflection off solar panels on their megastructures. This idea is based on the understanding that an alien civilisation would harness solar energy as stars are the most efficient producers of energy.

Caleb A. Scharf, an American astrobiologist, has come up with a unique way to quantify intelligent life, based on external *information* that a species generates, utilises, propagates and encodes what we call technology – everything from cave paintings and books to flash drives and cloud servers and the structures sustaining them. He calls it 'dataome'. In his 2021 book, *The Ascent of Information: Books, Bits, Genes, Machines, and Life's Unending Algorithm*, he says that our informational world, our dataome, is best thought of as a symbiotic entity to us (and life on Earth in general). It is genuinely another 'ome', not like microbiomes that exist in an intimate and inextricable relationship with multicellular life.

He suggests that technosignatures are a consequence of dataomes, just like biosignatures are a consequence of genomes. In brief, the arrival of a dataome on a world represents an origin event; therefore, the search for technosignatures is about the detection of extraterrestrial dataomes.

Let the search for dataome begin. What if privacy laws of aliens prevent them from disclosing their dataomes to anyone and are required to hide them under a cloud of technobabble?

If aliens are taking any interest in millions of words earthlings write about them every year, not to speak of streams of podcasts, documentaries and movies, they must also be burning with curiosity about inquisitive earthlings. If they focus their telescopes on us …

What would Earth look like to them?

Would they see the Great Wall of China? In 1938, well before the advent of space flights, American adventurer and intrepid traveller Richard Halliburton claimed in his book *Second Book of Marvels: The Orient* that the Great Wall is the only manufactured object visible to the unaided human eye from the Moon. This started the popular urban myth. The only thing you can see from the Moon, says NASA astronaut Alan Bean, is a beautiful sphere, mostly white (clouds), some blue (ocean), patches of yellow (deserts), and every once in a while, some green vegetation. From the window of a spacecraft in a low orbit of Earth you can see the Great Wall, the Great Pyramid of Giza and many other manufactured objects, if you know where to look and the weather is right. No artificial object is visible from the window of the International Space Station which orbits at about 360 kilometres up; however, you may spot big cities such as London, New York and Beijing in the daytime if you were good at your geography class.

The first-ever picture of Earth from Mars shows Earth blue and beautiful against the deep darkness of space. The picture was taken in 2003 by the Mars Global Surveyor spacecraft from a distance of 139 million kilometres (86 million miles). The most famous picture of Earth was taken in 1990 by the Voyager spacecraft from the edge of the solar system at a distance of 6.4 billion kilometres (4 million miles). In his 1994 book *Pale Blue Dot: A Vision of the Human Future in Space*, Carl Sagan used the picture as a metaphor for the insignificance of our world in comparison to the cosmos. From a distance of 100 or more light-years, the aliens would see the 'pale blue dot' in their sky only if they had unimaginably powerful optical telescopes.

CHAPTER 10

HYPOTHESES: GOOD, BAD AND MAD

The Incredible improbability of intelligent life

After life appeared on Earth, evolution took over. The main tenet of evolution, as proposed by Darwin in his monumental 1859 book, *On the Origin of Species by Means of Natural Selection*, is that all present-day species have evolved slowly, over millions of years, from simpler life forms through natural selection, which favours individuals that are better adapted for survival and reproduction.

Did you know about the reaction of the bishop's wife to the suggestion that man is derived from the apes? 'Let's hope that it is not true – or if it is, that it won't become generally known.' This is an example of the furore caused by Darwin's theory, which was in complete contrast with the religious notion of the creation of Earth, its life forms and the rest of the universe by a divine being. To the annoyance of the bishop's wife, the idea of evolution has become generally known. But it still continues to generate enormous social and scientific debate.

After about 3.8 billion years of evolution, humans are at its pinnacle. Intelligence has helped them to get there. Intelligence is difficult to define. Dolphins (capable of abstract communication), primates (can use simple tools) and African Grey parrots (can categorise objects), for example, are intelligent in their own ways, but they all lack the most important aspect of human intelligence: its creativity that has resulted in the development of technology. Is evolution of creative intelligence inevitable? Once life appears on another world, what are the odds of its eventually evolving into creative intelligence? We already have a tag for such intelligence – ETI (extraterrestrial intelligence) – and an expectation of its creativity: a seat at the control panel of a spacecraft (UFO fans' belief), or at least at the control panel of an interstellar radio station (SETI enthusiasts' hope).

Ernst Mayr, the renowned evolutionary biologist who died at 100 in 2005, was not optimistic about ETI. He notes that out of probably more than a billion species of animals that have arisen on Earth, only one succeeded in producing the kind of intelligence to establish a civilisation. Even this civilisation did not develop the capability of interstellar communication until a few decades ago. He stresses that the assumption that any intelligent extraterrestrial life must have a technology and mode of thinking like us was unbelievably naive.

In an innovative essay published in 1985, Mayr illustrates the incredible improbability of intelligent life ever to have evolved, even on Earth, by representing the history of life on Earth on the scale of a calendar year:

Origin of Earth	1 January
Life (prokaryotes)	27 February
Eukaryotes	4 September
Chordates	17 November
Vertebrates	21 November
Mammals	12 December
Primates	26 December
Anthropoids	30 December at 1:00 am
Hominids	31 December at 10:00 am
Humans	31 December at 11:56:30 pm

This calendar shows that humans appeared 3½ minutes before the year's end; they occupy only 0.025 per cent of the total of history of life on Earth.

The main thrust of Mayr's argument is as follows: What is most remarkable is that for about 2 billion years nothing very much spectacular happened as far as life on Earth was concerned. About 1.5 billion years ago a most remarkable event took place – the evolution of eukaryotes. (All organisms can be divided into prokaryotes (bacteria) and eukaryotes (all organisms except bacteria), the latter having a well-organised nucleus and chromosomes in each cell.) The most likely explanation for this event is the close association of two or more prokaryotes to create the first eukaryotes. After eukaryotes, an almost explosive innovative diversification took place. Within several hundred million years four new kingdoms evolved: the protists (one-celled animals and plants), fungi, plants and animals.

In none of these kingdoms, except that of the animals, Mayr says, was there even the beginning of any evolutionary trends towards intelligence. He illuminates the diversity of the animal kingdom and points out that, after more than 500 million years and numerous evolutionary lines, only primates showed true intelligence. It took another 25 million years and more evolutionary branches before one pathway eventually led to the rise of humans, less than one-third of a million years ago. There is no straight line from the origin of life to intelligent humans. 'The point I am making is the incredible improbability of genuine intelligence emerging,' he says. One inference of this point – that man is probably the most intelligent creature in the universe – should make the bishop's wife happy.

He points out another improbability: intelligence does not mean developing a technology capable of interstellar communication. Civilisations, as humans demonstrate, are fleeting moments in the history of an intelligent species. Even if two civilisations develop such a technology, it is necessary that they flourish simultaneously. He illustrates this improbability with a fable: 'Let me assume there is another high technology civilization in our galaxy. By some extraordinary instrumentation their inhabitants were able to discover the origin of the earth 4.5 billion years ago. At once they began to send signals to the earth and continued to do so for 4.5 billion years. Finally at the time of the birth of Christ they decided that they would terminate their program after another 1,900 years, if they had not received any answer by then. When they abandoned their program in 1900, they had proven to their own satisfaction that there was no other intelligent life in our galaxy.'

Mayr, however, does not want to 'deny categorically the possibility of extraterrestrial intelligence'. He just wants to claim that from an evolutionary biologist's point of view the probabilities are close to zero.

We live in a galactic zoo …

Extraterrestrial intelligent life is widespread. Their reluctance to interact with us can be explained by the hypothesis that they have set aside our planet as part of a wilderness area or zoo.

This is the controversial and demoralising hypothesis proposed by John A. Ball in 1973. The hypothesis is based on three premises:

1. *Whenever the conditions are such that life can exist and evolve, it will.* Bell believes that the discovery of primitive life on Mars or anywhere else would probably solve this question.

2. *There are many places where life can exist.* Since Ball proposed this hypothesis, the discovery of exoplanets supports this premise.

3. We are unaware of 'them'.

Ball considers the third premise extremely significant. We are not aware of them because they are deliberately avoiding us, and they have set aside our planet as a zoo or wilderness sanctuary. In a perfect zoo, animals would not interact with, and would be unaware of, their keepers. 'The hypothesis predicts that we shall never find them because they do not want to be found and they have the technological ability to ensure this,' he says.

He admits that the hypothesis is pessimistic and psychologically unpleasant: 'It would be more pleasant to believe that they want to talk with us, or that they would want to talk with us if they knew we are here.'

In a comment made in 1980 on the zoo hypothesis, Ball suggests that the civilisation that is number one exercises control and enforces the rules: 'They may keep us separate from our neighbours to prevent unfavourable or disastrous interaction.' Or perhaps they are waiting until the time is right to invite us to join the Galactic Club.

German astrophysicist Peter Ulmschneider also supports the zoo hypothesis. He says: 'While at first sight this idea appears quite extravagant, it makes considerable sense on closer inspection.' He suggests that our history would suffer drastic and fundamental changes – much more so than the history of the native Americans, annihilated when Columbus and Cortez arrived – if it came in contact with an alien advanced civilisation. 'It would be a catastrophe, a culture shock, and essentially an irresponsible act on behalf of the extraterrestrials,' he warns.

Ulmschneider, however, believes that a civilisation as capable of irresponsible acts as human civilisation has been in the past would not survive thousands, millions or billions of years without falling victim to the dangers of such behaviour. His conclusion: if highly advanced extraterrestrial civilisations exist, they have learned to act responsibly. This means 'under no circumstances will they disturb us or contact us, nor will they allow us to trace them by radio waves, through artefacts, or by direct contact'.

This rules out sending flying saucers to check on our well-being. Not a happy idea for UFO fans to entertain. Perhaps there is a sign somewhere at the perimeter of the solar system warning all trespassers: 'Wilderness Sanctuary: Don't Enter.' UFO fans can take solace in the knowledge that some cheeky alien teenagers do occasionally ignore this warning and fly their space buggies in our protected skies.

Ulmschneider is not so pessimistic about the future of the human race. He believes that one day we will have the knowledge to discover extraterrestrial intelligent societies in our own galaxy and other galaxies, but the possibility of ever directly interacting with them is very bleak. He is hopeful that, by that time, the basic idea behind the zoo hypothesis – that of acting responsibly – will become the guiding principle for our own behaviour. Moral: When we do escape from our zoo, we should become keepers of other zoos.

... or in a planetarium

We don't live in a zoo run by an extraterrestrial civilisation, but in an artificial universe, a kind of virtual reality planetarium designed to give us the illusion that the universe is empty. This hypothesis, published in 2001 by science fiction writer Stephen Baxter in *Journal of British Interplanetary Society*, maintains that this illusion cannot last forever as humans have now started to venture into space in earnest.

... or in a computer simulation

No, we don't live in a planetarium but a computer simulation. This idea was advanced in 2003 in *Philosophical Quarterly* by British philosopher Nick Bostrom who says, 'It follows that the belief that there is a significant chance one day we will become posthumans who run ancestor-simulation is false, unless we are currently living in a simulation.'

A technologically mature 'posthuman' civilisation would have enormous computing power, but the fraction of human-level civilisations that reach a posthuman stage is very close to zero. Bostrom assures us: 'Unless we are now living in a simulation, our descendants will almost certainly never run an ancestor-simulation.'

Computer simulation hypothesis has its detractors, but it also has high-powered supporters like Elon Musk. The billionaire entrepreneur said in 2018 that we are probably trapped in a Matrix-like pseudo existence. He was referring to the 1999 movie, The Matrix, in which humanity is trapped in a simulated reality. 'We are most likely in a simulation because we exist,' Musk said.

Whatever this gobbledegook means, ET, please don't be alarmed we will never entrap you in a computer simulation, planetarium or a zoo; you are welcome to visit us – not only on our movie screens but in your shiny IFOs.

... or on an extraordinarily rare Earth

Our planet is neither a zoo nor a planetarium of an extraterrestrial civilisation, but it's an extraordinarily rare planet. It has exactly the right location, moon, chemical composition and distance from the Sun to enable life to thrive. Its ideal location keeps it away from the rain of killer rocks, and thus safe from mass extinctions. Its large moon minimises changes in the planet's tilt, ensuring climate stability. It has enough carbon and other elements which help in the development of life and keep greenhouse conditions under control. Its right distance from the Sun ensures that water remains liquid, not vapour or ice.

These and many other unique features of Earth have convinced the University of Washington's Peter Ward, an authority on mass extinctions, and Donald Brownlee, an astronomer, to conclude that although microscopic life may be common on other worlds, complex life is rare. To them, even a flatworm is complex life. By this definition, the chances of existence of an extraterrestrial civilisation awash in technology are almost zero.

They argue in their book *Rare Earth: Why Complex Life is Uncommon in the Universe* (2000) that almost all environments in the universe are terrible for life, and Earth is probably the only Garden of Eden in the cosmos. They say that their rare Earth hypothesis, which has been called everything from 'simplistic' to 'brilliant' and 'courageous', is testable. They suggest two types of tests.

The first test is the search for microscopic life in other bodies in the solar system. This search can be done by specialised probes seeking life directly in samples. They suggest searching Mars, Jupiter's moons Europa and Ganymede, and Saturn's moon Titan for signs of alien microbes. The discovery of living micro-organisms or fossil evidence of micro-organisms will prove that life originates readily.

The second test involves the search for complex life forms. As there is no evidence of such life forms in the solar system, they suggest using powerful new telescopes to detect life on exoplanets.

As far as intelligent life is concerned, the two scientists support listening to alien radio signals. They advise against the more complex and costly exercise of beaming radio messages to nearby stars and are doubtful whether the search for extraterrestrial intelligence is an effective use of resources. 'If the rare Earth hypothesis is correct, then it is clearly a futile effort,' they say.

... or a mediocre planet?

At the 1971 'Communications with Extraterrestrial Intelligence' conference in Soviet Armenia, when a speaker droned on about his dubious theory that great scientific ideas are born during times when sunspots are active, Iosif Shklovsky whispered to Carl Sagan, who was sitting next to him, that this theory must have been conceived when sunspots were absent.

The witty Soviet astrophysicist died in 1985 at the age of 69, but his sharp wit can still be enjoyed in his riveting reminiscences, *Five Billion Vodka Bottles to the Moon: Tales of a Soviet Scientist*. He was one of the first eminent astrophysicists to take a serious interest in the possibility of extraterrestrial intelligence. In 1962 he published a book in Russian, *Universe, Life, Intelligence*, which became an instant bestseller in the Soviet Union (the first printing of 50,000 copies was sold out in a few hours).

When Shklovsky asked Sagan to edit the English translation of this book, Sagan added so many notes of his own that they doubled the size of the book. As a result, his name appears as a co-author of *Intelligent Life in the Universe* (1966). In the book, Sagan's text is distinguished with triangles. This was a priceless boon to me, notes Shklovsky, otherwise vigilant Soviet censors could have made things tough for me. The book became the bible of the new SETI movement.

In this book, Shklovsky and Sagan ask: Is by 'some poignant and unfathomable joke' ours the only civilisation in the universe? Their answer is that 'our surroundings are more or less typical of any other region of the universe'. A high percentage of stars have planets, most of these planets are Earth-like, life has developed on most such planets, and intelligent life has evolved on a large number of these planets. There's nothing special about Earth. It's a mediocre planet.

From this assumption of mediocrity, they conclude that 'all planets on which life has flourished for several billions of years have a high probability of the development of intelligence and technical civilization'. But they agree that this is at best a plausibility argument: 'we do not know the detailed factors involved in the development of intelligence and technical civilization.'

The idea of mediocrity was first proposed by German American astronomer Sebastian von Hoerner in the early 1960s. He said that the Sun is an average star, Earth an average planet, and that humankind has an average intelligence. In *Cosmic Search* (January 1977), the first magazine devoted to SETI and astrobiology, he explains his assumption of us being average from one example: 'We have only one single case and that is us. And the question is, can we do statistics when the sample number N equals one? The answer is "yes", if you know the rules.

'If you have only one example to go by, with N equals one, then you do have an estimator for the average, and that is the one case you have. This would mean we should assume that we are average. On the other side with N equal to one, we do not have an estimator for the mean error. In plain English this means that the assumption that we are average has the highest probability of being right, but we have not the slightest idea of how wrong it is.'

He concludes that if we generalise our own case, we should expect a very large number of extraterrestrial civilisations with whom we could talk.

The selfish biocosm

Life not only plays a role in shaping our planet, it also plays a role in shaping the universe. The cosmos is remarkably hospitable to life. It's a living creature, a biocosm. It's so finely tuned to life that if any of its fundamental physical laws and constants – from the 'just right' strength of the Big Bang that produced it to the relative strengths of gravity, electromagnetism and the so-called strong and weak nuclear forces – were slightly different, there wouldn't be any life. The laws and constants are intelligently designed to yield life and even more competent intelligence. Why is the universe so life-friendly?

Life and intelligence have not emerged in a series of random events, but essentially are hard-wired into the cycle of cosmic creation, evolution, death and rebirth. This hard-wiring was done by super-intelligent extraterrestrials to give birth to life and intelligence. Life and intelligence, in turn, are the means used by the universe to reproduce itself by creating 'baby universes' which themselves, in turn, are able to create more 'baby universes'.

No, this is not from the blurb of a new science fiction book, but a summary of a new hypothesis, called Selfish Biocosm, which has won praise, if not approval, from many scientists. James Gardner, an American complexity theorist, science essayist and lawyer, presents it in his book *Biocosm*). The essence of the hypothesis, according to Gardner, which he himself labels 'speculative', is that 'the universe we are privileged to inhabit is literally in the process of transforming itself from inanimate to animate matter'.

He agrees that, to be acceptable as a genuine scientific hypothesis, it should be testable. He presents four possible tests: a) the discovery of extraterrestrial intelligence would prove the hypothesis; b) the proof of evolution of sentience in non-primate species such as dolphins would support the hypothesis by indicating that the emergence of intelligence is a relatively robust, species-neutral phenomenon; c) the creation of artificial life and its ability to evolve would support the hypothesis; and d) the emergence of trans-human machine intelligence, the combined human and machine intelligence, is necessary for the feats of cosmological replication.

It will all happen when life on Earth becomes super-intelligent like the extraterrestrial super-intelligent beings that created it. Gardner believes that humans and their probable progeny are the evolutionary predecessors of the supremely evolved intelligence that will emerge in the distant future. At the end of the cosmic evolutionary cycle, a super-intelligent entity is sufficiently advanced to be capable of engineering, or re-engineering, the basic laws and constants of physics. Let's find out whether the present most intelligent life form on Earth will live long enough to become one day the super-intelligent 'architects of the universe'.

They're already in your backyard

Physicist and science writer Paul Davies does not deny the existence of extraterrestrial intelligence. He speculates that an alien civilisation could use nanotechnology to build miniature space probes, not necessarily self-replicating probes, perhaps no bigger than your palm.

In 1959 the Nobel Prize-winning physicist Richard Feynman gave a classic talk, 'There's Plenty of Room at the Bottom', in which he said that 'the principles of physics, as far as I can see, do not speak against the possibility of manoeuvring things atom by atom' – or building molecular-sized machines. The advance of nanotechnology – the technology that deals with objects smaller than 100 nanometres (a nanometre is a millionth of a millimetre), especially manipulation of individual atoms and molecules – has fulfilled his prophecy.

'The tiny probes I'm talking about will be so inconspicuous that it's no surprise that we haven't come across one,' Davies says. 'It's not the sort of thing that you're going to trip over in your backyard. So, if that is the way technology develops, namely, smaller, faster, cheaper, and if other civilisations have gone this route, then we could be surrounded by surveillance devices.'

In fact, he believes in the possibility of a probe resting on our own moon, having arrived at some time in our prehistory and remained to monitor our planet. American theoretical physicist Michio Kaku supports this idea. 'Personally, and this is a view shared by Paul Davies, a friend of mine, we believe that we have already been visited; and chances are they are on the Moon,' he says. The aliens are watching you. Let's us find out how 'civilised' they are.

We lag behind in the civilisation stakes

The famous Russian astrophysicist Nikolai Kardashev was a doctoral student of Shklovsky at the Sternberg Astronomical Institute in the early 1960s. In his reminiscences, Shklovsky is somewhat dismissive of Kardashev's and other scientists' optimism for extraterrestrial intelligence. He believes that 'the problem of extraterrestrial civilizations is in essence and effect *complex*'.

In spite of his teacher's misgivings, Kardashev was an earlier advocate of the idea that some alien civilisations may be billions of years ahead of us. In 1964, two years after obtaining his doctorate at the age of 30, he classified these advanced civilisations by a level of technological development that shows their use of energy resources for interstellar communications purposes. A Type I civilisation has enough energy resources to muster the equivalent to the entire present energy consumption of the planet Earth, a Type II civilisation the energy output of its own star, and a Type III the total energy output of its galaxy. We are not yet a Type I civilisation, according to this classification. Carl Sagan has classed our present civilisation as something like Type 0.7. At present we consume about one million billionth of the Sun's total energy received by Earth. Freeman Dyson has estimated that within 200 years or so we should attain Type I status, growing a modest rate of 1 per cent per year.

Kaku has enhanced Kardashev's concept of classification of civilisations. According to him, a Type I civilisation can control the weather, earthquakes, tsunamis and volcanoes. Once they exhaust the power of a planet, they go to their nearby star. A Type II civilisation uses so much energy that 'their planet may glow like a Christmas tree ornament'. The only threat to their existence is a supernova explosion which may kill all life forms. A Type III is truly immortal, truly galactic in power and scope. It even has the power to change the evolution of a star before it explodes into a supernova.

On Kaku's scale, we qualify only for a Type 0 status. But he is optimistic, as he can see the seed of a Type I civilisation: 'We see the beginning of a planetary language (English), a planetary communication system (the Internet), a planetary economy (the forging of the European Union), and even the beginnings of a planetary culture (via mass media, TV, rock music, and Hollywood films).' If you speak English, work in Germany, read *Il Globo* on the internet, listen to the Rolling Stones on your iPhone and watch Netflix series and Hollywood movies on your iPad, you'll soon be allowed to wear a citizenship ring of the Shire of Greater Earth, a truly Type I civilisation. Citizenship has its privileges. The power to control the weather, earthquakes, tsunamis and volcanoes – on any Xbox in the universe.

Are alien visits possible technically?

How long will it take aliens to travel to Earth from their home, say, in Zeta Reticuli? Thirty-nine years if they travel at the speed of light. Einstein has banned travel at this speed, at least for earthlings. His special theory of relativity says that the mass of a moving object increases as its speed increases. At the speed of light, which is about 300,000 kilometres(186,000 miles) per second, the mass becomes infinite, and therefore nothing can move faster than light.

But this Einsteinian stricture hasn't stopped scientists speculating about faster-than-light travel through 'wormholes' in space-time. Interestingly, the idea of wormholes was made possible by Einstein's theory of general relativity in 1916. However, scientists didn't take much interest in them until Carl Sagan employed a wormhole in *Contact* to transport Dr Allie Arroway to the centre of the galaxy and back.

Our universe appears as space-time: three dimensions in space (up-down, left-right, and forward-backward) and a fourth dimension known as time. Wormholes are 'tunnels' between two different places in space and time. Imagine a worm crawling on a flat leaf from point A near one edge to point B near the other edge. If the leaf is curled up, the worm will take less space (and time) to crawl through the 'shortcut tunnel' between points A and B. The idea seems simple in theory, but the wormholes scientists talk about are extremely unstable, short-lived and narrow (not even wide enough for a proton to pass through). There is another small matter:

the nearest wormhole gateway is through the massive black hole at the centre of the galaxy. If aliens – which may be thousands, millions or billions of years ahead of us – have found ways of breaking our laws of physics, a new Wormhole Tunnel Express station may soon open near your bed or car. You have been warned.

As far as we earthlings are concerned, wormholes are like the rabbit-hole in *Alice in Wonderland*. However, Michio Kaku believes that faster-than-light travel is probably attainable in 100,000 to 1 million years. 'When I look at the age of the universe, I see that we've attained technology in the blink of an eye. There's plenty of time.'

Some scientists are impatient. These scientists – for whom 'wildly optimistic dreamers' is virtually a job description, according to the *New York Times* – are not willing to wait for a million years to colonise other worlds. They pin their hopes on what they call multi-generational space travel in a rocketless spaceship. The spaceship will be equipped with ultra-thin, ultra-large sails which will be propelled by a beam of laser light from Earth or high-energy particles from the sun-soaked planet Mercury. The spaceship will travel at about 30 per cent of the speed of light, or about 88,000 kilometres per second. Compare it with the speed of Voyager, the fastest human-made object in space, which is about 15 kilometres per second. The spaceship would take about 50 years to reach the nearest star, Alpha Centauri, about 4.4 light-years away. Who would want to spend their next 50 years in a spaceship to reach Alpha Centauri to meet alien rocks, or even alien rock stars?

One of these 'wildly optimistic dreamers' is John Moore, an anthropologist at the University of Florida. He dreams of ships powered by multi-generational crews – people who man the ship with their children, grandchildren and descendants through six, eight or ten generations. Moore's computer simulations show that a founding crew of 80 to 100 people would provide enough genetic variability to make it viable for more than a thousand years.

If intelligent extraterrestrials would have thought of this *Star Trek* plan, they would be here; and Fermi would not have asked his question.

Rasmus Bjørk of the Niels Bohr Institute in Copenhagen has taken a down-to-earth approach to 'investigate the possible use of space probes to explore the Milky Way, as a means to both finding life elsewhere in Galaxy and as finding an answer to the Fermi paradox'. He says that the use of interstellar probes is the only way to perform detailed investigations of exoplanets believed to harbour life. His calculations show that eight probes – travelling at a tenth of the speed of light and each capable of launching eight sub-probes – would take 100,000 years to explore 40,000 star systems in the habitable zone of the galaxy. He agrees that exploring the galaxy by space probes is 'horribly slow'.

Intelligent extraterrestrials seem to agree with Bjørk; and that's why Fermi asked the question: If they are there, why aren't they here?

CHAPTER 11

HELLO ET!

Say hello in prime numbers

SETI researchers do not expect to hear a hello from an ET from their speakers, but they hope to see their computers recording an intelligent pictorial or mathematical message. A series of short and longer transmissions representing dots and dashes could be used to send a pictorial message. A mathematical message could be sent as a prime number or as Fibonacci numbers: 1, 1, 2, 3, 5, 8, 13, 21, 34, 55, etc. (each successive term is the sum of the preceding two).

Fibonacci numbers have many interesting universal properties. For example, the ratio of successive terms (larger to smaller: 1/1, 2/1, 3/2, 5/3, 8/5 …) approaches the number 1.618. This ratio is known as the golden ratio and is denoted by the Greek letter phi (Φ). Phi was known to ancient Greeks, and Greek architects used the ratio 1:Φ as part of their designs, the most famous of which is the Parthenon in Athens. Curiously, phi also appears in the natural world. Flowers often have a Fibonacci number of petals (look at the arrangement of florets on a cauliflower). The seeds on a sunflower are arranged in two sets of spirals. The ratio of the number of seeds in the two spirals is phi – and so is the ratio of your height and the distance of your belly button from the tip of your feet. Do the aliens have belly buttons?

Mathematicians are probably not interested in the answer, but they believe that mathematics is universal and that any intelligent aliens will understand the language of mathematics. 'Are the mathematics and the mathematical descriptions of physical phenomena used by an alien technological civilization on another planet likely to resemble what human beings come up with on Earth, even if the notation is different?' asks Murray Gell-Mann, the American theoretical physicist who introduced the concept of quarks.

He recalls an anecdote from his term as a visiting professor at the Collège de France in the 1960s. He was discussing Project Ozma with someone, and how communication with the aliens might take place: 'I suggested we might try beep, beep-beep, beep-beep-beep, etc. to indicate the numbers 1, 2, 3, and so forth, and then perhaps 1, 2, 3, … 42, 44, … 60, 62, … 92, for the atomic numbers of the 90 chemical elements that are stable 1 to 92 except for 43 and 61.' The person replied: 'Wait, that is absurd. Those numbers up to 92 would mean nothing to such aliens … Why, if they have 90 stable chemical elements as we do, then they must also have the Eiffel Tower and Brigitte Bardot.'

Gell-Mann replies by saying that the aliens will arrive at the same physical laws 'even if each of them has seven tentacles, 13 sense organs, and a brain shaped like a pretzel'. He agrees that their mathematical notation is very unlikely to resemble ours. Most scientists also believe that the fundamental laws of science provide a common 'language' throughout the universe, since they hold true in any region of it.

Leopold Kronecker, a 19th-century German mathematician, had a different view: 'The integers were created by God; all else is the work of man.'

In the 21st century, Indian philosopher Sundar Sarukkai notes that even if numbers or counting may have a common genesis, mathematics on other worlds may differ considerably from ours. He provides two arguments:

There are no unique mathematical descriptions of physical phenomena. We extract physical concepts from these phenomena, which are then mathematically described. And the way we extract physical concepts from the world depends to a great extent on our physical and mental capacities. For example, if our eyes were sufficiently tuned to detect that space is curved, we could have known it much before Einstein! Thus, the physical description we extract from the world to which we match certain mathematical ideas depends to a great extent on our capacities of perception.

The mathematics known to us arises from our interaction with our physical world. Thus, it depends not only on the nature of our physical world but also on the nature of our perception. We see the world in discrete terms, such as seeing objects as discrete entities. Moreover, the mathematics we create depends on the structure of our body and of our mind. There is no human mathematics without writing and there is no writing without fingers that do the writing. Would intelligent aliens who have no writing develop mathematics as we know it? I think the answer is no. For creatures which have different bodies and minds, there is no conceivable way that they would have the same mathematics as we do.

Although mathematics is not universal in the way scientists think of it, he believes, it does not preclude the possibility of using mathematics for communicating with aliens. 'This is primarily because of the capacity for mathematics to be taken into the folds of translation in very interesting ways,' he says.

Jason Wright, an astrophysicist at Pennsylvania State University, disagrees. He believes that our mathematical practices, such as attachment to prime numbers, might prove to be idiosyncratic. 'We are looking for kindred spirits that will find interesting what we've found interesting,' he says. Aliens may turn out to be more creative than our fantasies. We have to find ways to transcend our limited ideas.

ET, we will soon call you in a new language

Douglas Vakoch, president of METI (Messaging Extraterrestrial Intelligence) International, a San Francisco-based non-profit dedicated to messaging alien civilisations, argues that we need to find a unique way to speak across the stars than we have used in the past.

Messages based on maths and science are not the best way to contact aliens. It might make more sense to focus on what mathematics and language have in common. We must draw on some of the cognitive processes that undergird such different human creations as mathematics and language.

The radio technology that enables interstellar communication is less than a century old – a blink in the in a universe nearly 14 billon years old. 'As infants in the cosmos, we should begin with baby steps; a few key concepts that make communication possible in the first place,' Vakoch says.

'The man who speaks for Earth', as his speakers agency bills Vakoch, may find it easy to say hello to ET, but the hard part is to have a conversation with aliens whose bodies and minds are likely to be quite different from our own. If aliens are smarter than we are, they will find ways to communicate with us. Maybe telepathically with their wide eyes, as alien abductees believe!

A singing astrogram

Humanity's first deliberate message to other civilisations was sent from the Arecibo radio telescope in 1974. It was beamed towards Messier 13, a group of some 300,000 stars in the constellation Hercules 25,000 light-years away. Carl Sagan estimated that there is 'about one in a two chance of there being a civilization in Messier 13'. The message was transmitted in 169 seconds at a radio frequency of 2,380 megahertz (wavelength 12.6 centimetres). The strong signals containing the message, which were repeated many times, will be detectable by Arecibo-sized radio telescopes throughout the galaxy.

The message consists of 179 consecutive characters written in binary numbers, one of the simplest number systems with only two symbols, 0 and 1. In the actual transmission, 0 and 1 were represented by one of the two specific radio frequencies, and the characters were sent sequentially by shifting between these two nearby frequencies.

The message has a series of 1679 bits, which can be broken down into a grid, 23 characters by 73, which can be used to build a bitmap image. The number 1679 is equal to the product of the prime numbers 23 and 73, and it is believed that these will help the aliens to arrive at the fact that this grid is the only correct way to interpret the message.

The following figure shows the 1974 Arecibo message to the stars in binary numbers (left) and in a bitmap image (right). (Images: National Astronomy and Ionosphere Center, Cornell University)

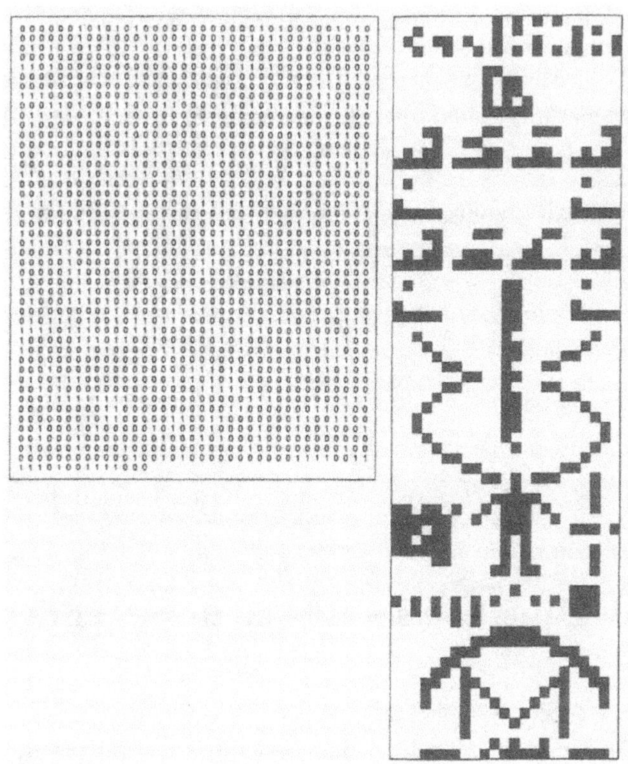

The opening line of the message, right to left, describes numbers 1 to 10 in binary form. This 'lesson' will help the receiver in deciphering the message. The next three lines show the atomic numbers of five chemical elements – hydrogen, carbon, nitrogen, oxygen and phosphorus – that are important in living cells.

Next, the message uses this information to describe the molecular components of DNA. The double spiral below shows the DNA molecules; and at the spiral's central core is the number 4 billion – the number of characters in the genetic code. The DNA is followed by the crude sketch of a human (it looks like the legendary Australian bushranger Ned Kelly in gumboots, according to a wit), establishing the connection between the DNA and the intelligent creature. To the right of the human is a line extending from the head to the feet and accompanied by the number 14. This means that the height of the human is 176 centimetres, 14 times 12.6 centimetres – the unit of wavelength that is used to transmit the message. To the left of the human is the human population of Earth (4 billion).

The next line shows a sketch of the planets of the solar system, with the Sun at the right and Earth displaced towards the human figure, suggesting that the third planet is the home of the creatures who sent the astrogram. The message is signed by the Arecibo telescope with its diameter of 305 metres (1000 feet) shown as a multiple of 12.6 centimetres (nearly 5 inches).

The message was developed by Frank Drake and his colleagues at Cornell University's National Astronomy and Ionosphere Center. The message was 'market tested' to see how easy it was to decipher. 'But a spot check in the *Nature* office revealed no intelligences (native or alien) that could decipher it,' reported the journal on 24 January 1975. 'Are interstellar IQ tests culture fair?' the journal wondered.

No codes, just the sounds of the seventies

Identical engraved plaques attached to Pioneers 10 and 11 spacecraft, launched in 1972 and 1973 respectively, were definitely culture fair. Each plaque measures 15 by 22 centimetres (5.9 by 8.7 inches) and is made of gold-anodised aluminium. These cosmic greeting cards were designed by Carl Sagan and Frank Drake and drawn by artist Linda Salzman Sagan.

On the plaque, a nude human male and female stand before an outline of the spacecraft coming from the third of nine planets around a star (see figure below). The star's location is shown with respect to 14 pulsars and the centre of the galaxy as the radial pattern of dashed lines at left centre. The precise periods of the pulsars are shown in binary numbers to allow them to be identified. The heights of man and woman are shown with respect to the spacecraft, and the height of the female is also given as a binary number eight on the far right of the plaque.

Units of time and distances are expressed relative to the wavelength of 21 centimetres emitted by a hydrogen atom. This element is illustrated in the left-hand corner in schematic form showing the hyperfine period of neutral hydrogen (a fraction of a nanosecond). If alien physicists can understand this information, they can unscramble the message. For example, they can work out that 8 times 21 centimetres is the height of the female earthling.

According to Sagan and Drake, the plaques provide sufficient information to our star in 250 billion stars, and one year, 1970, in 10 billion years. 'These plaques are destined to be the longest-lived works of humankind. They will survive virtually unchanged for hundreds of millions, perhaps billions, of years in space.'

Contact with Pioneer 11 was lost in 1995 and Pioneer 10's last signal was received in 2003. Both spacecrafts are still travelling as ghost ships in our galaxy.

Voyager 1 and 2, launched in 1977, each carry a 30-centimetre (12-inch) gold-plated copper phonograph record containing 115 images and greetings spoken in 55 languages, selected to portray the diversity of life and culture on Earth. Each record is packed in an aluminium jacket together with a stylus and cartridge. It's designed to play images and sounds at $16^2/_3$ revolutions per minute. The record's image and sound are encoded in analogue, which has been superseded by digital coding. 'With hopeless gallantry, we built these messages to last a billion years with playback equipment that was obsolete before it got to Neptune,' laments E.C. Krupp of Griffith Observatory Los Angeles. Both Voyager spacecraft are now at the edge of the solar system and on track to a faraway radio station to play golden sounds of the seventies.

The following figure shows the gold-anodised plaque affixed to Pioneer 10 and 11 spacecraft (launched in 1972 and 1973, respectively). (Image: NASA)

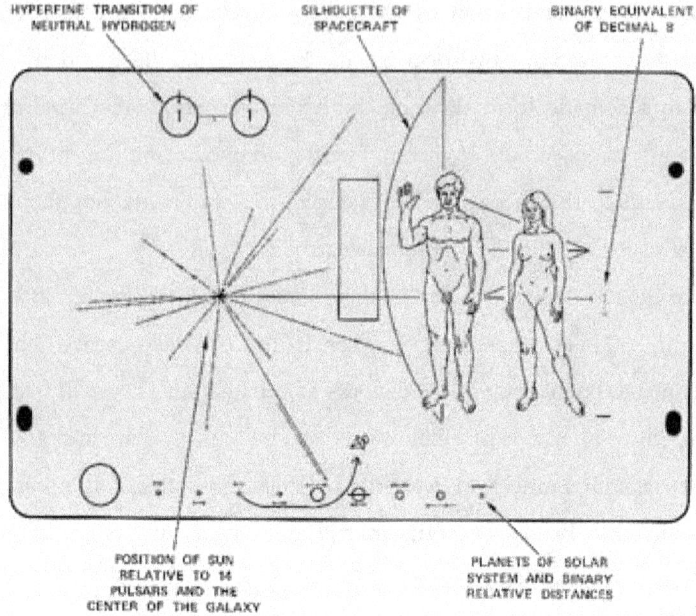

Message in light fantastic

Why, despite decades of listening, haven't we picked up any obviously intelligent radio signals?

Perhaps no civilisation is continuously broadcasting a powerful 'beacon' signal from a planet around any of the nearest 1,000 or so Sun-like stars. If such signals existed, they would have been picked up by powerful searches undertaken by SETI researchers in the past decade.

Or, we are searching such a small range of frequencies that it's like poking around the edges of the 'cosmic haystack' for the 'needle' of an alien signal. This haystack is estimated to have 100 trillion radio channels, sky directions and other parameters. 'Earthlings are just getting into the game,' remarks Dan Werthimer, an American SETI astronomer. 'We can rule out hardly anything.'

Or, ETs are quite imaginative, and they may have moved to technologies far superior to radio signals. Such as lasers.

American physicist Charles Townes, who shared the 1964 Nobel Prize for the invention of lasers, realised around the time that Drake aimed his telescope at distant stars that ET civilisations could just as easily exploit the optical and infrared portion of the spectrum as the radio portion. Decades passed before laser technology had advanced to the point where powerful lasers capable of sending interstellar messages could be made.

Lasers produce intense light of usually very pure frequency or wavelength, which is emitted in an extremely tight beam. This beam can be easily focused on a target as small as one millionth of a millimetre. High frequencies produced by lasers, several hundred million megahertz, give them enough bandwidth for a very high rate of transmission of information. Thus, the laser has a major advantage over the millions of radio channels available for broadcasting: it can transmit a whole movie, if the aliens are into movies, in seconds – much better than simply asking 'How are you, earthlings?' by a radio wave. They offer another advantage. If we do receive a laser signal, it will be much easier to locate the source, as lasers are uni-directional.

SETI enthusiasts began searching for laser signals in the early 1990s, but they have not yet beamed any laser messages to other planets. Paul Horowitz, a Harvard University physicist, is a veteran of laser searches. He notes that lasers have now grown so powerful that a laser beam could be 5,000 times brighter than the Sun. Similarly, optical telescopes fitted with photodetectors have now become so powerful that they can record extraterrestrial pulses of laser light only a few billionths of a second (nanosecond) in duration. Using a 1.5-metre telescope at Oak Ridge Observatory, Horowitz and his colleagues have made more than 16,000 searches for nanosecond pulses from 13,000 Sun-like stars. No ET seems to be beaming a copy of *Encyclopaedia Galactica* in our direction.

Townes now believes that flashes of light from planets around stars within 50 light-years could even grow bright enough for the naked eye to see. When it does happen, all you have to do to search for ETs is to relax in a lounge chair and look up at the sky.

Message in a bottle

If aliens want to contact us, they are not likely to send their *Encyclopaedia Galactica* 'we are here' message by a pulse of radio waves or lasers, but in an old-fashioned way – by a written note (in a perfumed envelope, we hope). After analysing the energy efficiency of long-distance communication, American researchers Christopher Rose and Gregory Wright suggest that 'our initial contact with extraterrestrial civilizations may be more likely to occur through physical artefacts – essentially a message in a bottle – than via electromagnetic communication'. They explain their calculations with the concrete example of the Voyager spacecraft, which is carrying a recording containing about 1 billion bits of information. If it were carrying three DVDs with 100 billion bits of information, they calculate, it would be a more energy-efficient way of sending this amount of data to someone 200 light-years away than an Arecibo-to-Arecibo radio communication.

Unless the message is short or a prompt reply is desired, sending a message inscribed on some material requires less energy per bit of transmitted information than sending it by radio waves, the researchers say. 'If energy is what you care about, it's tremendously more efficient to toss a rock.' Beams of radiation spread out as they travel through space, diluting the signal and wasting energy. An inscribed physical object, on the other hand, is not 'diluted' as it travels through space. Radio announcements last momentarily unless repeated continuously. Physical objects last forever, almost.

A message in a bottle could be waiting for us in our own planetary backyard, like the obelisk left on the Moon by aliens in Arthur C. Clarke's *2001: A Space Odyssey*. Any artefacts from aliens are likely to be orbiting the Sun or a planet, or resting somewhere on a planet, moon or asteroid. Rose and Wright advise that 'carefully searching our own planetary backyard may be as likely to reveal evidence of extraterrestrial civilizations as studying stars through telescope.'

The *New York Times* is among those who are sceptical of receiving a bottled message from aliens: 'The only hitch is that such a message, travelling at a crawl by space standards, might have been sent 30 million years ago, and who knows whether the senders would have moved on or died off in the eons since. Best not to give up on electromagnetic messages entirely.' ET, please phone us.

Sorry, the number is not yet connected

For interstellar communication, messages should travel as fast as possible, preferably at the speed of light. That's why we expect extraterrestrials to communicate by radio or other electromagnetic waves. But some interstellar civilisation may be using neutrinos instead of electromagnetic waves for communication. Neutrinos are elementary particles with no charge and are nearly massless. They travel at or near the speed of light. They are everywhere – trillions of them pass through our bodies each second – but they cannot be seen and rarely interact with matter. They are the most elusive of all the elementary particles we know, yet scientists have found ways to detect these ghost-like particles.

A massive neutrino telescope is being built at the South Pole to trap neutrinos from deep space. These originate in black holes or crashing galaxies which give them enough energy to travel billions of light-years. The telescope, dubbed IceCube, will sample neutrinos from the sky in the northern hemisphere. Earth will act as a filter to exclude neutrinos originating from the Sun. When completed by the end of the decade, the telescope will house 60 basketball-sized optical detectors in each of 70 2.4-kilometre-deep holes drilled into the Antarctic ice. These 4,200 optical detectors will occupy one cubic kilometre of ice, hence the name IceCube. The detectors will record tell-tale signatures of neutrinos when they pass through Earth from the northern hemisphere and collide with other atoms. This rare collision produces another elementary particle called a muon. The muon leaves a trail of blue light in its wake which allows scientists to trace its direction back to the point of origin of the neutrino. IceCube may even one day eavesdrop on alien teens talking on their nifty neutrino mobiles.

Walter Simmons and his colleagues at the University of Hawaii in 1994 were the first to suggest that extraterrestrial civilisations might be using neutrinos to broadcast information throughout the galaxy. They came to this conclusion after reasoning that any advanced civilisation is bound to have exacting time standards. A civilisation that has spread through our galaxy might send 'timing pulses' over interstellar distances in order to synchronise clocks on different worlds. The decay of a particle known as a Z boson into a neutrino and an anti-neutrino takes just one thousand billion billionth of a second, making it the fastest process known in physics. Neutrinos have other advantages over photons (which make up electromagnetic waves): they are not blocked by interstellar dust or dispersed by ionised gas and are brighter than photons. A very bright source of neutrinos could transmit precise timing pulses across many thousands of light-years. These neutrino signals will be quite distinctive, as they will come from a specific direction in the sky and have a very well-defined energy.

No 18th-century scientist could have dreamt of sending sounds and pictures around the planet by electromagnetic waves. What about a civilisation trying to communicate with us via a medium as yet undreamt of? Our neutrino communication lines are now open, but they might be sending 'smoke signals' we are not yet ready to decipher.

Look for 'smoke signals' in space

These 'smoke signals' might be in the form of giant solar sails in space, if we believe Luc Arnold of the Observatoire de Haute-Provence in France. 'Artificial structures may be the best way for an advanced extraterrestrial civilisation to signal its presence to an emerging technology like ours,' he says. These planet-sized structures could be like lightweight solar sails or other very low-density structures specially built for the purpose of interstellar communication.

When a planet crosses in front of its star as viewed by an observer, the event is called a transit. Transits produce a small and periodic change in the parent star's brightness. All transits of the same planet produce the same change in brightness and last the same amount of time, thus providing a highly accurate method of detection. If planet-sized artificial objects exist around other stars, they would always transit in front of their star for a given observer in space. These objects can be spotted by a new generation of space-based telescopes such as NASA's Kepler Mission. Kepler was specifically designed to detect Earth-sized or smaller planets in our galaxy by studying their transit. Kepler is now out of fuel and its transmitter has been shut down.

After studying several simulated artificial transits, Arnold has determined the characteristic transit signals of differently shaped objects. A Jupiter-sized triangle, for example, will produce specific transit signals which are different from the transit signals of a Jupiter-sized planet. Multiple objects produce remarkable transit signals because of their 'on again – off again' nature of light. Such an observation would clearly show that the transiting objects are artificial. The best way to visualise this is to imagine a flashlight moving behind a lowered window blind.

Arnold says that a civilisation would rather build a series of small objects to generate multiple transits rather than a large single object for saying 'Hi, we're here' to the whole galaxy. He imagines, for example, that if 11 objects were orbiting a star in groups of 1, 2, 3 and 5 – the first five prime numbers – the mathematical pattern of the dimming of the star's light would clearly represent a message from the aliens.

'Transit of artificial objects also could be a means for interstellar communication *from* Earth in future,' he says. His advice to future generations: to have in mind, at the proper time, the potential of Earth-sized artificial structures in orbit around the Sun to say hello to intelligent extraterrestrials. They may not have the technology to decipher messages sent by radio waves, lasers or neutrinos, but they can't miss 'smoke signals' in the sky.

The message is in our genes

Forget radio waves, lasers, neutrinos or giant structures around stars – the most likely place to find an alien message is in our genes. Paul Davies believes that the aliens have left their message in the junk DNA in our genes. The term 'junk DNA' is used for those portions of DNA which appear to have no specific function.

Scientists have recently discovered large sequences of junk DNA which have not changed over the last 75 million years of evolution.

A DNA sequence can contain more than a million base pairs, which is enough for a decent-sized book. The DNA need not to be the last word from ET, says Davies. 'Rather, it could tell us how to download the entire contents of *Encyclopaedia Galactica* by conventional radio or optical techniques.'

Why can't the aliens broadcast the encyclopaedia now? Davies says that it's inconceivable that ET would beam signals at our planet continuously for untold aeons merely in the hope that one day intelligent beings might evolve and decide to turn a radio telescope in their direction. Nor does he entertain the idea of the aliens leaving an obelisk at our doorstep, because such an object would be subject to the vagaries of tectonic activity, glaciations and other natural turmoil.

From his point of view, 'a better solution would be a legion of small, cheap, self-replicating machines that can keep editing and copying the information and perpetuate themselves over immense durations in the face of unseen environmental hazards'. Nothing beats the living cell. He speculates that ET might have inserted a message into the genomes of terrestrial organisms, perhaps by delivering carefully crafted viruses in tiny space probes to infect host cells with message-laden DNA.

To read this cosmic greeting card, Davies suggests displaying a sequence of junk DNA bases as an array of pixels on a computer screen. Patterns that stand out, such as prime numbers, would be a clincher.

ET, please call Earth in Lincos

Lincos – short for the Latin *lingua cosmica* – is a universal language developed in 1960 by Dutch mathematician Hans Freudenthal. It's not a spoken language, but a complex mathematical scheme for use in interstellar radio communication with extraterrestrials.

Freudenthal suggested that first we should 'communicate facts which may be supposed to be known to the receiver'. Obviously, these facts must be concepts that are universal. Humans learned to count before they learned to write; mathematical concepts are believed to be universal. The first lesson in Lincos starts with a simple pattern of radio pulses to establish symbols for natural numbers in binary notation and basic arithmetical processes: addition, subtraction, multiplication and division. Each symbol is explained by symbols that came before it, so you don't have to know anything except pure mathematical concepts to understand it. Freudenthal believed that 'decoding Lincos would be an easy job'. However, ETs must have the technology to receive radio signals and measure their frequency. Lincos phonemes that make up words are transmitted as radio signals of varying duration and frequency.

The first interstellar message in Lincos was broadcast in 1999 from a 70-metre radio telescope in Evpatoria, a small town in Ukraine. It was aimed at stars 51 to 71 light-years away in a region of sky called the Summer Triangle. The signal was 100,000 times stronger than a TV broadcast. The 400,000-bit message was much longer than the 1,679-bit Arecibo message broadcast in 1974.

The message was developed by Canadian physicists Yvan Dutil and Stéphane Dumas. SETI is a hobby for both of them, and the message was developed as a private project. They believe in active SETI – the active search for extraterrestrials. 'In classic SETI you listen as hard as you can, in active SETI you shout as loud as you can,' says Dutil. 'By the way, to be honest, active SETI is the true classic SETI, since it was first proposed by the German scientist Gauss at the end of the 19th century.'

In 2003 Dutil and Dumas improved their 1999 'shout', and it was broadcast on the same day from radio telescopes in Evpatoria and Roswell, New Mexico. Unlike a pure Lincos message, their message also uses graphics. The 1999 message was an interstellar fax of 23 pages; the 2003 message is one awfully long page. In fact, it's a lesson in mathematics, physics, biology and cosmology. It will be difficult for visually challenged aliens to decipher it, as some form of sight is a prerequisite. We can only hope that if they can develop the technology to receive it, they may have the ability to 'see' it.

Will aliens be able to understand it? Easily, if they have an Alan Turing among them. The message will decipher itself, as it also teaches those who receive it how to crack the code. 'There are numerous ways of failing not that many of succeeding,' admits Dumas. 'There is absolutely no guarantees it will work.' Dumas also points out that Lincos was written more for earthlings than ET.

The following figure shows a geometry lesson for aliens. Euclid might not have understood it, but the aliens will. The top section introduces radius and area. In the middle, pi is displayed with 1,241 billion digits. The bottom section shows the graphical representation of Pythagoras' theorem. (Image: Yvan Dutil and Stephane Dumas)

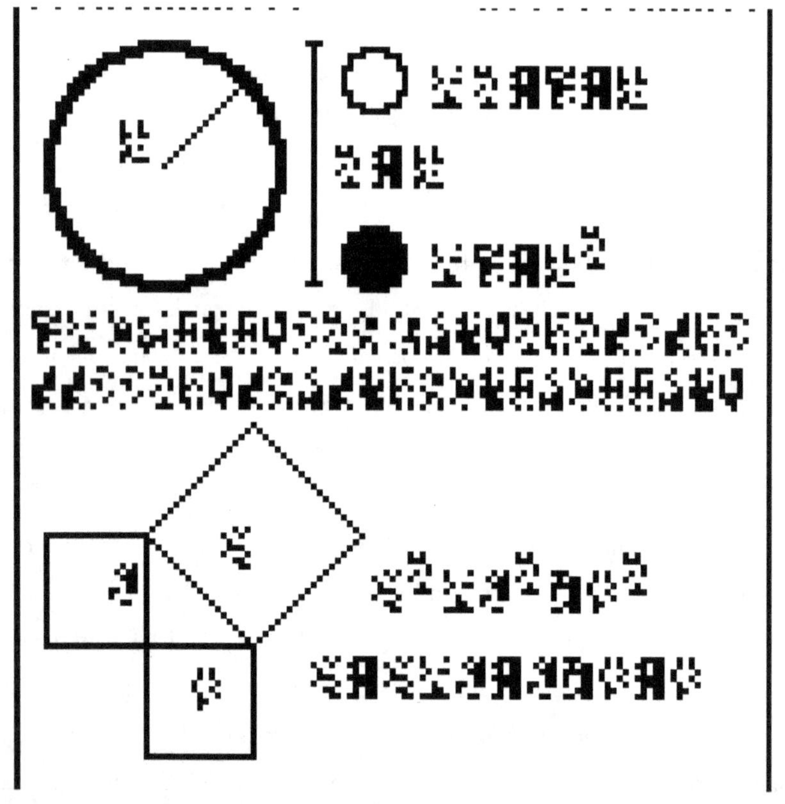

What if they want to remain hidden?

Suppose ETs have received the Evpatoria message, and they want to send a thank-you card to earthlings for teaching them Pythagoras' theorem and other such earthly things. But they don't want their message traced back to them; they want to remain hidden.

Walter Simmons and his University of Hawaii colleague Sandip Pakvasa believe that they know how ETs can do it. They say that in cryptography a random signal is added to a message to make it indecipherable by anyone other than the sender and the intended recipient. Both the sender and the recipient know the random signal. When the signal is subtracted from the message, it becomes decipherable. ETs can achieve the same effect by splitting their message into two parts and sending them in opposite directions to two mirrors located far from the sender's planet. The mirrors reflect the signals to the intended receiver. The message is revealed when the two parts are combined.

The researchers then solve the problem of how the two parts cannot be decrypted individually, either before or after they are combined and read. They suggest that if the message is tiny enough to fit on a microscopic particle, Heisenberg's uncertainty principle would make it impossible to read the message and the direction simultaneously. The principle says that it's impossible to determine exactly the position and momentum of a particle simultaneously. By reading the message, all the information about its origin will be lost.

There is another problem: how two signals can carry an image without leaving any trace of its presence in either signal. The researchers expect a quantum effect known as 'entangled particles' to help ETs. This effect allows a pair of photons to be connected at a distance and to influence each other instantaneously, whether they are in the same room or at opposite ends of the galaxy. The events affecting one photon will also modify the other photon. Einstein never liked this 'telepathic connection' between particles. He called it 'spooky action at a distance'. The Hawaii duo expects quantum entangled photons to unite to form the microscopic image of the alien message. Scientists have recently shown that entangled particles can indeed be used to carry an image.

ETs may be using this strategy to send interstellar messages to other Galactic Club members, if for some unknown reason they do not want to reveal their presence to us. We do not have the technology to look for such alien signals today, Simmons and Pakvasa say, but we may have it in a decade or so.

This is Earth calling, again

But aliens living on exoplanets ignore us; we've had no replies. Here're some other messages, still travelling or lost in space:

Teen-Age Message: In 2001 a group of Russian teens, taking part directly and via the internet, composed a message that included a 15-minute concert of electronic music. The message was transmitted from Evpatoria Space Center to six nearby Sun-like stars which have own planets. We don't know whether alien teens on any planets of these planets will wait until 2047 to listen to the seven melodies of the concert. The message also has text of greetings to extraterrestrials both in Russian and English, and an image glossary of basic terrestrial concepts.

A whole website: In 2005 the classified listing website Craiglist was transmitted into space by private company, Deep Space Communications Network. The message was sent into open space, rather to specific stars. It is unlikely it will be picked up by aliens keen to sell their old TVs to earthlings.

Goodwill messages from ordinary earthlings: In 2009, Australia's Cosmos magazine collected goodwill messages from its readers and transmitted the best ones to the planet Gliese 581d, a planet orbiting a red dwarf star 20 light-years away. The messages were sent from the Canberra Deep Space Communications Complex. They should arrive in 2028.

CHAPTER 12

WHISPERS FROM ET

A missed phone call from ET?

It came from outer space and lasted only 72 seconds. On the computer printout it simply appeared as 6EQUJ5 – a code that revealed it was a strong, intermittent radio signal confined to a narrow band of frequency. It was so unusual that it caused an excited astronomer to scrawl 'Wow!' in the margin of the printout, a label that is inextricably linked to it.

What was it? A message from an alien intelligence? A momentary hiccup from a cosmic event, or a polluting burp from a terrestrial transmission? Nearly five decades on, no one knows what really created the signal, and the debate continues.

The Big Ear radio telescope at the then Ohio State University Radio Observatory had been involved in the search for extraterrestrial intelligence since 1973. On the night of 15 August 1977 its 79-metre dish was tuned to 1,420 megahertz, the frequency of hydrogen atoms.

Interstellar space is filled with hydrogen atoms, at a density of about one atom per cubic centimetre. The individual atoms chirp at a frequency of 1,420 megahertz, like a miniature radio station. The chorus of these countless atoms can easily be heard by any radio telescope. As hydrogen is the most common element in the universe, many people believe that the aliens might choose this frequency to broadcast their presence to us. Even if the aliens do not use this frequency, they are likely to send a narrow-band signal, which has a precise frequency and a longer range. A broad-band signal, on the other hand, is like a noise that covers a wide range of frequencies. Stars and many other cosmic objects emit broad-band noise. Radio stations broadcast narrow-band signals, but the hiss between stations is broad-band noise. The terrestrial intelligence advises that an alien intelligence would not want their message to be mistaken as ordinary star noise.

For this reason, the Big Ear had been converted from measuring the location and strength of wide-band signals to narrow-band signals. In a radio telescope, signals received from a large area are focused by the dish to a receiver that amplifies the fluctuating voltages of the signals. You can't just plug your headphones into the receiver and listen to the chatter, natural or alien made. Even if you could, all you would hear is a hissing sound. The Big Ear used a computer to record the signals. The floor-standing desk-size IBM 1130 used by the observatory was equipped with 16 kilobytes of RAM and a megabyte of storage on a magnetic disk. This stone-age storage capacity meant that the daily observations had to be printed out and examined by volunteers, as the observatory did not have any funding for the programme.

Jerry Ehman, a professor at Ohio State University, took the responsibility for this task. A few days after 15 August, he began his routine review of the printouts, not expecting to find anything unusual. As he worked through his way through the reams of paper for the night of 15 August, he was astonished to see the string of numbers and characters 6EQUJ5 on the printout. It represented a burst of radio waves, like a thunderclap in the middle of a piece of quiet music. Ehman immediately recognised it as a narrow-band signal confined to a frequency around 1,420 megahertz. Without thinking, he wrote 'Wow!' and circled the string 6EQUJ5 on the printout. 'It was the most significant thing we had seen,' recalls Ehman. For a month following the discovery, he and his colleagues looked for the signal again at least 50 times but found nothing.

Each character of the alphanumeric code used by the observatory's computer represented the strength of a received signal over 12 seconds. The code 6EQUJ5 revealed that the 'Wow' signal rose and fell over the course of 72 seconds:

6	6
E	14
Q	26
U	30
J	19
5	5

(The numerals in the right-hand column show approximate signal-to-noise ratios; that is, how many times the signal is stronger than the background noise. For example, the strongest signal received, U, was about 30 times stronger than the background noise. Most of the background noise is generated within the receiver itself; some noise comes from buildings, trees, grass and other surroundings, and the celestial sky.)

The following figure shows a section of the computer printout showing Jerry Ehman's 'Wow!' and the circling of 6EQUJ5 (Image: Jerry Ehman, North American Astrophysical Observatory, the continuing organisation of the Big Ear Radio Observatory)

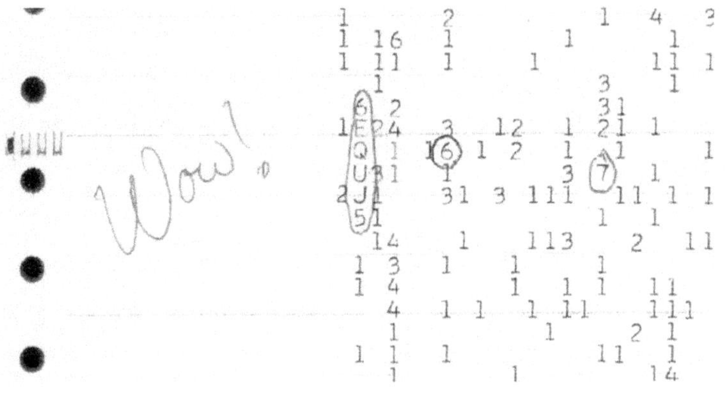

The fact that the signal rose and fell over the course of 72 seconds is intriguing. The Big Ear was a fixed telescope (it was demolished in 1998), and the Earth's daily spin allowed it to pick up cosmic signals from a tiny angular section of the heavens. The string 6EQUJ5 shows that as the radio source passed by, its intensity rose as the Earth's spin brought it within the telescope's range, reached a peak in the centre, and then faded away. For the Big Ear, this rise and fall should last 72 seconds, and that's what happened. If the signal were from a terrestrial source, it would suddenly flood the telescope and then switch off after some time.

Under certain conditions, a terrestrial signal reflected off a piece of space debris could appear as if it were coming from a point source like the Wow signal. 'I now place a low probability on this alternative,' says Ehman. A terrestrial signal also scores low probability for another reason: radio transmissions in the frequency band around 1,420 megahertz are prohibited by international agreement.

Did the Wow signal include a message? The signal was unmodulated. Unlike modulated AM or FM radio signals (in which characteristics of the signals are changed or modulated to include data), it was a blast of radio noise. However, it is still possible that there could be modulation at frequencies that were not within the range of the observatory's detection equipment. 'We would not have seen that modulation, and hence we could say that modulation is within the realm of possibility,' remarks Ehman.

Ehman and his colleagues also looked at other possible sources of the signal:

Planets: None of the planets was close to the signal source position. Planets do generate some radio waves, but they have much broader frequencies.

Asteroids: Asteroids are essentially small planets. None of the larger asteroids was in the vicinity.

Satellites: An investigation of the orbits of all known satellites showed that none was within the telescope's range. Satellites do not use the protected frequency band of 1,420 megahertz.

Aircraft: Aircraft also do not transmit in the protected band. An aircraft will also show a significant motion with respect to the stars, which would cause the pattern of signal to be very different from that expected from a point source.

Spacecraft: Spacecraft are also prohibited from transmitting in the protected band; none was within the telescope's range.

Interstellar scintillation: We see the stars twinkling because their light is scattered unevenly by the Earth's atmosphere. Similarly, when light and radio waves travel through interstellar space, they also 'twinkle'. It is possible that a radio signal could become stronger as it passes through interstellar space. If it did occur to the Wow signal, it still points to a signal coming from a source many light-years away from us. 'This gives more support for the hypothesis of a signal of an extraterrestrial origin,' says Ehman.

Gravitational lensing: Gravity can bend light and radio waves. Did the gravity of an intervening star focus a much weaker signal into the stronger Wow signal? Ehman rules out this explanation on the ground that gravitational lensing usually lasts many days or months depending on the motion of the source.

So, the evidence stacks up in favour of an alien source. Not quite. The Big Ear used two funnel-shaped metal structures called horns, situated side by side, to collect radio waves focused by the dish. The path of the radio waves collected by a horn is called a beam. As Earth rotates, any cosmic radio source would be seen first in beam one (for 72 seconds) and then – about three minutes later – in beam two (also 72 seconds). Ehman's computer printout showed it only on one beam, instead of the two beams expected. The computer was not programmed to identify whether the signal came from the first or the second beam. 'Suspicious and disheartening,' remarks Seth Shostak of SETI Institute, a California-based non-profit organisation that conducts research on life in the universe.

Perhaps the aliens turned off their transmitter after three minutes and went on holiday. Was the signal a wish-you-were-here postcard? No one knows. Unlike alien astronomers, terrestrial astronomers are workaholics. From observatories around the world, they have performed more than 100 searches of the same region of the sky. No report yet of an astronomer running naked through the street, shouting 'Wow! Wow!'

Robert Gray, an independent SETI researcher from Chicago, has been trying to solve the riddle of the Wow signal for more than a decade. In 2001, using a radio telescope 100 times more sensitive than the Big Ear, he looked for a radio source – artificial or natural – at the position that Wow came from, but found nothing. In a recent survey, with Simon Ellingsen of the University of Tasmania, he looked at the possibility of a periodic source to explain the single-beam detection of the Wow signal. Pulsars, stars that emit rapid pulses of radio waves, are familiar examples of cosmic periodic sources of radio waves. Using a 26-metre radio telescope, Gray and Ellingsen made six 14-hour observations of the Wow spot in the sky. Their conclusion: 'No signals resembling the Ohio State Wow were detected.'

Calculations show that the Wow signal came from the direction of the constellation of Sagittarius; the nearest star in the constellation is 220 light-years away. If someone were sending a message from such a faraway star, they would need a large 300-metre dish and a gigantic 2,000-megahertz transmitter. 'That's daunting, but not altogether impossible,' assures Paul Shuch, an American SETI enthusiast. Even if we believe Shuch, the message might not have been meant for us. It might have been a message from an alien civilisation to another alien civilisation. Did the Big Ear inadvertently eavesdrop on an interstellar phone conversation?

Ehman believes that even if the Wow signal were an alien signal, they would do it far more than once: 'We should have seen it again when we looked for it 50 times.' Shostak agrees: 'So until and unless the cosmic beep measured in Ohio is found again, the Wow signal will remain What signal.'

A tiny blue spark heralds the radio age

James Clerk Maxwell, whose scientific achievement is comparable with that of Newton or Einstein, showed so little promise at school that his classmates nicknamed him 'Dafty'. But when his father took him to a science lecture, he soon became interested. A few years later he sent his first scientific paper to the Royal Society of Edinburgh. The paper presented a novel technique for drawing a perfect oval and was so well done that no one could believe it was written by a 14-year-old boy.

In 1864, when he was 33 years old, Maxwell read to the Royal Society of London his most important paper, showing a connection between electricity and magnetism. The paper also included his four equations – now famous as Maxwell's equations – that express mathematically the way in which electricity and magnetism behave. Though the equations predicted the existence of electromagnetic waves, which travel at the speed of light and consist of electric and magnetic fields vibrating in harmony in directions at right angles to each other, Maxwell never proved their existence.

In 1886, seven years after Maxwell's death, Heinrich Hertz, a newly married young professor at the Technische Hochschule in Karlsruhe, Germany, set out to find the elusive 'wireless waves', as they were then known. He modified an induction coil to generate sparks across a gap between two brass balls. In those days it was a common set-up for demonstrating electric discharge. A few metres away from the induction coil he placed a loop of wire connected to two brass balls separated by a tiny gap. When he passed the discharge through the induction coil, he was startled to see a tiny spark in the loop of wire a short distance away. His wife was with him to witness the first radio transmission in history: one wave emitted at one point had been received at another. No music, no talkback shows, just a tiny blue spark and an entry in Hertz's diary for 1 November 1886: 'Vertical electric vibrations, in wires stretched in straight lines, discovered; wavelength, 3 metres.'

If Maxwell, who loved reading and writing poems, were alive, perhaps he would have rewritten his poem, 'Valentine from a Telegraph Clerk ♂ to a Telegraph Clerk ♀', for the new age of interstellar communication. An excerpt:

'O tell me, when along the line

From my full heart the message flows,

What currents are induced in thine?

One click from thee will end my woes'.

Through many an Ohm the Weber flew,

And clicked the answer back to me,

'I am thy Farad, staunch and true,

Charged to a Volt with love for thee'.

(In Maxwell's time, the term 'Weber' was used for Ampere and 'Farad' for

Coulomb.)

The tiny blue spark proved that electromagnetic waves do exist. A year later, Hertz was able to measure the speed of these waves and to show that the speed was the same as that of light. His further experiments showed that electromagnetic waves could be refracted, reflected and polarised in the manner of light.

Hertz never realised the importance of his discovery. When he demonstrated his experiment to his students, someone asked what the discovery could be used for. 'Nothing, I guess,' replied Hertz. It was left to the Italian physicist Guglielmo Marconi and others to develop technology for the practical use of Hertzian waves.

We are now familiar with all types of electromagnetic waves or radiation that make up the complete electromagnetic spectrum. These waves are streams of photons travelling in a wave-like motion with the speed of light. Photons are bundles of energy. They have no mass, and their energy increases with their frequency. The only difference between various types of electromagnetic waves is the energy of the photons. In order of increasing energies, electromagnetic waves are called radio waves, microwaves, infrared, visible light, ultraviolet, X-rays and gamma rays. This order also shows increasing frequencies or decreasing wavelengths. Wavelength, the distance between two successive crests or troughs of a wave, is measured in metres. Frequency, the number of crests – or complete cycles – that pass a given point in one second, is measured in hertz (Hz), a unit named in Hertz's honour.

Our noisy cosmic neighbours

While Hertz was looking at ways of sending 'wireless waves' through air, British physicist Oliver Lodge was experimenting with transmitting these waves along electric wires. Hertz's discovery overshadowed his work. Nevertheless, the feisty physicist continued his research on electromagnetic waves and in 1894 suggested that if heat, light and radio waves were different manifestations of electromagnetic waves, radio waves might also be coming from the Sun to Earth. In 1897 he attempted some experiments to pick up radio waves from the Sun by blocking heat and light by an opaque substance. He admits the failure of his experiments in his 1897 book, *Signalling across Space without Wires*: 'There were evidently too many terrestrial sources of disturbances in a city like Liverpool to make the experiment feasible.' Lodge's failure to communicate with the Sun (curiously, he spent his later life experimenting with the possibility of communicating with the dead) meant that the world had to wait another 36 years to hear the news of the discovery of extraterrestrial radio waves and the birth of radio astronomy.

The news first appeared in the *New York Times* of 5 May 1933 and filled an entire front-page column. The report, headlined, New Radio Waves Traced to the Centre of the Galaxy, denied that they were signals from extraterrestrials: 'There is no indication of any kind ... that these galactic radio waves constitute some kind of interstellar signalling, or that they are the result of some form of intelligence striving for intra-galactic communication.'

On 27 April 1933, Karl Jansky, a 37-year-old engineer from the Bell Telephone Laboratories, read a paper entitled 'Electrical Disturbances of Extraterrestrial Origin' to a small audience at a meeting of the International Scientific Union in Washington, DC. The paper, which received a lukewarm reception, is now viewed as the beginning of radio astronomy. A few days later, Bell Labs' spin doctors decided to send out a press release. Media organisations all over the world picked up the news and Jansky became an instant celebrity. On 15 May, Jansky appeared on an NBC radio programme. Listeners throughout the United States heard the hiss of the stars live by a telephone hook-up to the receiver of Jansky's primitive radio telescope at Holmdel in New Jersey. The next day, the *Times* headlined: Radio waves heard from remote space … after travelling 30,000 light-years. The report described the hiss as 'sounding like steam escaping from a radiator'.

When Jansky joined the Bell Labs in 1928 as a radio engineer, the recently opened New York to London radio telephone service was plagued with intermittent noisy crackling of static interference caused by various sources such as electrical equipment and thunderstorms. Bell Labs' research engineers were interested in reducing noise levels in telephone conversations. Jansky's job was to record the intensity of this interference with a radio receiver connected to an antenna – a long array of metal pipes – mounted on four Ford Model T wheels. An electric motor moved the wheels on a 3-metre sprocket to point the antenna to any part of the sky. During one of his experiments, Jansky heard a steady hissing sound at a frequency of 20.5 megahertz (or a wavelength of 14.6 metres), which was very different from the intermittent crackling of the static. He had the insight to realise that the sound was not of terrestrial origin. He also ruled out the Sun as the possible source. He soon discovered that the signals became louder whenever the antenna pointed towards the constellation of Sagittarius. The young engineer had accidentally opened a new window to the universe. But it took scientists a long time to understand the significance of Jansky's discovery and look through this new window. Even Jansky did not pursue this new field further.

No story of radio astronomy is complete without a mention of the pioneering work of Grote Reber. He was a 22-year-old student at the Illinois Institute of Technology in Chicago when he heard about Jansky's discovery. He was so stimulated by it that he applied to almost all major research centres for astronomy in America to continue the research. No one showed any interest. Radio astronomy was a strange beast to them. In 1937, Reber decided to build a visionary 9-metre bowl-shaped antenna in his backyard. Using this first true radio telescope, for nearly a decade he conducted an extensive survey of the sky and produced the first maps of radio sources in the galaxy. The world was now ready to hear from our noisy cosmic neighbours.

Listening to the 'songs' of the cosmos

Because the wavelength of radio waves is much larger than the wavelength of light – some are longer than a kilometre – radio telescopes are much larger than optical telescopes. Radio telescopes are metal dishes that focus radio waves to a point. The spherical dish of the world's largest telescope, at Arecibo in Puerto Rico, is 305 metres in diameter, 51 metres deep and covers an area equal to seven football fields. Suspended 137 metres above the dish is a platform that houses a set of movable antennae and a highly sensitive radio receiver. To make radio images sharper, radio astronomers also combine many smaller receiving dishes into an array. The Very Large Array Radio Telescope in New Mexico consists of 27 antennae arranged in a huge 'Y' shape up to 36 kilometres across. This arrangement works as a radio telescope with a virtual dish 36 kilometres in diameter.

Most astronomical objects – planets, asteroids, comets, stars, giant clouds of gas and dust, black holes and galaxies – emit radio waves. According to a NASA website, if we had radio antennae instead of ears, we would hear a remarkable symphony of noises coming from our own planet. These noises are around us all the time. Scientists call them 'tweaks' (they sound like a quick musical ricochet), 'whistlers' (sound like a musical descending tone) and 'sferics' (short for 'atmospherics', sound like twigs snapping or bacon frying). These terrestrial radio waves can be converted into audible sound by a very low frequency (VLF) radio receiver.

Jupiter is another source of exotic sounds. As one would expect from the biggest planet in the solar system, these sounds are much louder than terrestrial sounds. Two types of Jovian 'songs' can be heard by a radio receiver with a special antenna in the shortwave bands between 15 and 40 megahertz: 'L-burst' (sounds like ocean waves catching on a distant beach) and 'S-burst' (a staccato of rapid popping sounds with a beat that reminds one of woodpeckers).

The Sun is the most intense source of radio waves in the solar system. Artificial radio signals from Earth come next. Whenever there are sunspots or solar flares, the intensity of solar radio waves increases. These waves can be heard directly, particularly at dawn, on a shortwave radio receiver at frequencies close to 20 megahertz.

Outside the solar system, the sources of radio waves include clouds of gas and dust, stars of various kinds such as pulsars (rapidly rotating neutron stars), quasars (short for 'quasi-stellar radio sources', galaxies during early stages of their life), black holes (a region of space-time from which nothing can escape, not even light) and galaxies (usually elliptical galaxies). The source of Jansky's radio hiss was most likely a supermassive black hole at the centre of our galaxy.

The idea that launched a thousand searches

To extraterrestrial 'civilizations with scientific interests and with technical possibilities much greater than those now available to us', ... 'our Sun must appear as a likely site for the evolution of new society. It is highly probable that for a long time they have been expecting the development of science near the Sun. We shall assume that long ago they established a channel of communication that would one day become known to us, and they look forward patiently to answering signals from the Sun which would make known to them that a new society has entered the community of intelligence.'

This is an extract from the first two paragraphs of a seminal paper published in 1959 by Philip Morrison and his Cornell University colleague Giuseppe Cocconi. The paper was the first to suggest the idea of searching for extraterrestrial intelligence by radio astronomy.

'What sort of a channel would it be?' they ask. Their answer: electromagnetic waves. They say that 1,420 megahertz is the ideal frequency: 'It is reasonable to expect that sensitive receivers for this frequency will be made at an early stage of the development of radioastronomy. That would be the expectation of the operators of the assumed source, and the present state of terrestrial instruments indeed justifies the expectation.'

They end their paper with the exhortation that these speculations should not be confined to the domain of science fiction:

'Few will deny the profound importance, practical and philosophical, which the detection of interstellar communications would have. We therefore feel that a discriminating search for signals deserves a considerable effort. The probability of success is difficult to estimate; but if we never search, the chance of success is zero.'

A year later, in 1960, 29-year-old Frank Drake made the first real attempt to eavesdrop on radio broadcasts from extraterrestrials. He aimed a 26-metre radio telescope at the stars Tau Ceti and Epsilon Eridani, some 11 light-years away. The telescope's receiver was tuned to a frequency of 1,420 megahertz. On the very first day of the project, recalls Drake, 'WHAM! A burst of noise shot out of the loudspeaker, the chart recorder started banging off the scale, and we were all jumping at once, wild with excitement.' The strong, clear signal was precisely 'what you would expect from an extraterrestrial intelligence trying to attract attention – as though they'd just been waiting for us to tune in'. The signal was probably caused by a secret US military experiment. The erroneous idea that Frank's team made a secret discovery of the project and its 'association' with the US military is what conspiracy theories are made of. The signal has earned a place in the folklore of UFO fans.

The project – dubbed Project Ozma after the queen in *The Wizard of Oz* (Oz was a place 'very far away, difficult to reach, and populated by strange and exotic beings') – lasted only two months. Except for that short signal, all they heard from the loudspeaker was static, and the chart recorded nothing but formless wiggles. After nearly half a century, numerous other SETI projects have not done any better.

Waiting at the waterhole to say hello to ET

Radio waves are not absorbed by interstellar gas and dust, which block visible light. They require only modest power and inexpensive equipment to transmit and are easy to detect. These two qualities make them the method of choice for any interstellar civilisation wanting to make itself known to other interstellar civilisations.

Within the radio spectrum, SETI researchers prefer to look for signals in the so-called 'waterholes'. The first waterhole is between 1,420 and 1,638 megahertz (wavelengths 21 and 18 centimetres). The first frequency is associated with hydrogen (H) and the second with hydroxyl radical (OH). When H and OH combine, they form water (H_2O) – hence the name waterhole. Extraterrestrial life forms based on water, like us, are likely to use these frequencies. The second waterhole is around 22,000 megahertz (wavelength 14 millimetres), the frequency associated with water. The third waterhole is the cosmic microwave background – frequency 150,000 megahertz or wavelength 2 millimetres – which is explained in the next section.

When searching for intelligent radio signals, SETI researchers face two choices: should they make 'targeted' searches or 'wide-sky surveys'? In a targeted search, researchers examine individual nearby stars with high-sensitivity radio telescopes. In a wide-sky survey, a larger swathe of the sky is scanned with lower sensitivity. Seth Shostak does not consider the two strategies as complementary. He would like to abandon star-by-star targeting in favour of scanning the areas of sky that have a greater number of stars, even if most of them are very far away. 'Unless ETs truly infest the stars like flies (very unlikely),' he says, 'the first signals we can detect will come from very rare, very powerful transmitters very far away.'

If aliens are not transmitting signals deliberately, we may have to eavesdrop. For eavesdropping, TV signals are ideal. The Earth's atmosphere absorbs most radio signals, but higher-frequency TV signals travel past the horizon into space and can then continue for distances of hundreds of light-years. It is possible that right now someone 54 or so light-years away is watching the coronation of Queen Elizabeth II, broadcast live on BBC on 2 June 1953. Similarly, if we ever receive a leaked extraterrestrial TV signal, it would have been transmitted decades ago, even centuries or millennia ago.

Intelligent aliens looking for TV signals as signs of life in our solar system must hurry up, as 'Earth is going to disappear very soon'. That's the advice from Frank Drake. He's worried that within 50 years or so, cable and direct-broadcast satellites will replace terrestrial TV transmission. Cables do not leak any signals into space, and the power of satellite broadcast is very low, about 20 watts per channel, all efficiently directed straight down towards Earth's surface. For humans, TV signals 'are the strongest signs of our existence', Drake says. Most SETI researchers also focus on eavesdropping alien signals produced for a planet's domestic use, such as our own TV, FM radio and radar signals, but inadvertently leaked out into space. Such signals are weak and difficult to spot. A beacon – a coded repetitive signal beamed to attract attention – from an alien civilisation is perhaps SETI researchers' best hope.

George W. Swenson, Jr., an American electrical engineer, is not optimistic about radio searches. He has calculated that the power required for sending a signal in all directions to a distance of 100 light-years is 5.8 million billion watts, which is 7,000 times the total electricity-generating capacity of the United States. There are only about 1,000 stars within 100 light-years, and the kind of systems that would be needed to mount a realistic project to beam a signal to a large number of them are probably beyond the resources of our society. Even if contact could somehow be made, the time delay before a response to a message could be received might very well stretch into many centuries. 'This is clearly a project for many generations in succession,' he warns. 'In all likelihood, it will require an enduring organization based on immutable dogma – like one of the world's major religions.'

Shostak has an immutable belief in intelligent aliens and their society's capacity to generate radio signals. He also has faith in terrestrial technology. He believes that advances in computer processing power and radio telescope technology will ensure that we detect alien signals within the next two decades. Shostak has called upon Drake's equation and Moore's law for his prediction. Using Drake's equation, he has arrived at an estimate of between 10,000 and 1 million alien radio transmitters broadcasting in our galaxy. He believes that Moore's law – that computer processing power doubles every 18 months – will hold true for the next ten years (as it has done for the past 40 years). For the following ten years, he takes a conservative estimate of a doubling of computer processing power every 36 months. This increased processing power will allow the analysing of radio signals from all stars between 200 and 1,000 light-years away. We have about two decades to think about what we are going to say in our own message, and then a wait for at least two centuries to hear the reply from our friendly aliens. But there's already a message for us in the sky, if we can read it.

A hidden message in the sky

Our universe began when an unimaginably dense and unimaginably hot speck of matter exploded spontaneously. The newly born universe was so hot that electrons and nuclei could not combine to form matter. The free-moving electrons scattered the photons, making the universe opaque. After 380,000 years, the universe cooled to about 4,500 degrees Celsius and electrons and nuclei could combine to form the first hydrogen atoms. The universe now became transparent, and photons were free to escape as high-energy gamma rays. As the universe continued cooling and expanding, the wavelength of the radiation stretched. It changed from short wavelength gamma rays to longer wavelength X-rays, ultraviolet rays, visible light, and after 13.7 billion years into microwaves. The remnant radiation, usually referred to as the cosmic microwave background, has a frequency of 150,000 megahertz (wavelength 2 millimetres); and its temperature is very cold, about –270 degrees Celsius (3 degrees above absolute zero). This leftover warmth from the primeval fireball fills the universe today. If we could see it, the entire sky would glow with astonishingly uniform brightness in all directions.

To American theoretical physicists Stephen Hsu and Anthony Zee, the cosmic microwave background is like a giant billboard in the sky, visible to every technologically advanced civilisation in the universe: 'Anybody can see it, regardless of whether you have five heads and three eyes, even if your biology is not based on carbon.' They believe that this billboard is the ideal place if the universe's creator wanted to leave a single universal message for its future inhabitants.

Cosmic microwave background has tiny temperature variations that could serve as a string of zeros and ones for a binary message. In 2001 NASA launched its WMAP satellite to measure the temperature variations of the cosmic microwave background, and it has since produced maps of the microwave sky. These maps will help scientists answer fundamental questions about the origin and fate of the universe. The graphs of WMAP measurements of temperature differences between pairs of points on the sky plotted against their angular separations create curves with series of peaks and troughs. These peaks and troughs so far have not revealed any secret code. It may be 20 or 30 years before satellites are sensitive enough to collect accurate data to determine whether Hsu and Zee's idea has any merit or if it's just plain wacky.

The physicists estimate that the maximum amount of information that could be written in a microwave message is about 100,000 bits. The first few bits must specify the key; otherwise, no one will ever be able to understand it, even if they stumble upon it. What do Hsu and Zee expect to find in the message? The Holy Grail of physics, of course, the elusive theory of everything. They believe that the equations describing the fundamental laws of physics will make sense to all civilised life forms in the universe and are compact enough to fit into a small chunk of bits.

Hsu and Zee urge that when more accurate cosmic microwave background data becomes available, it should be analysed for possible patterns. To them, looking for a message in the sky is more fun than searching for messages from aliens.

CHAPTER 13

WAITING FOR ULTIMATE ANSWERS

Hello Earth!

Imagine this: Scientists have received an alien radio signal. After a painstaking process of checking and verification, they believe that the signal is really from an extraterrestrial intelligence. What happens now? A five-minute hypothetical on this Earth-shattering discovery.

Importance: 'It would certainly be the greatest discovery of all time, eclipsing the findings of Newton, Darwin and Einstein combined,' comments Paul Davies. 'The knowledge that we are not alone would affect people's psyche and totally transform our world view.'

'The discovery would remove the final cloak of superiority,' adds Seth Shostak: 'We would know that we are neither culturally nor intellectually supreme, but simply one society among many. We would no longer be special.'

Secrecy: Will there be an *X-Files*-type cover-up or pressure from the authorities to classify information? No, emphasises Shostak: 'The real problem is not that an alien transmission would, or could, be kept secret. Rather, scientists worry that a promising signal would generate so much excitement that it would be announced prematurely, only to prove later to be of earthly origin.'

Interpretation: Once the signal has been proved to be from an intelligent civilisation, scientists' top priority would be to interpret the message. If they are not successful in interpreting it without any ambiguity, American author John Casti (*Paradigms Regained*, 2000) is worried that it may 'maximise the opportunity for malicious exploitation, as the public would be distracted from the scientific attempts to figure out what the message really means by the sideshow of events centring on those claim to know the meaning or claim that the SETI community and/or government knows the meaning and will not divulge it'. This scenario, he says, presents enormous opportunities for rumour, exploitation and fear.

Impact: The impact of the message on our society would depend upon its correct interpretation and its intention: whether the message is a simple greeting, 'Hello Earth!'; instructions to download the latest *Proxima Centauri Idol* show and such other goodies by radio waves; a command to pack up all our uranium in large crates and leave them in the middle of the Sahara desert so that they can be picked up by an alien spaceship; or a declaration of invasion in the style of *The War of the Worlds*. Most likely it would be a passive greeting directed at our solar system, not specifically at Earth, suggesting that, like us, the sending civilisation is scanning the galaxy for intelligent life.

Sociological studies suggest that the message would lead to confusion and excitement, with a desire by individuals to 'know more', but little panic or hysteria, says a publication from the SETI Institute. For some sceptics, the news could be disturbing.

The proof of the existence of intelligent life elsewhere in the universe might reinforce the religious beliefs of some people. Still, it might create problems for those who take their religious texts literally. Some religious groups might reject the idea altogether, but major religions are expected to cope well with this discovery, as they have with other scientific discoveries.

Reply: There is no rush to reply. The message might have taken hundreds of light-years – 100 light-years is the most optimistic estimate by SETI researchers – to travel to Earth. Our reply would take centuries, even millennia, to reach them. The sending civilisation might even become extinct (and their planet turned into a *Planet of the Apes*) by the time the reply reaches them.

As there is no international treaty defining the consultation process, the discovery might lead to political bickering among nations.

Should we reply?: George Wald, the famous biologist who won the 1967 Nobel

Prize for physiology or medicine and who believed that life probably originated on other worlds, was worried about humankind's self-esteem when he remarked in 1974: 'I can conceive of no nightmare so terrifying as establishing such communication with a so-called superior (or if you wish, advanced) technology in outer space.'

Zdeněk Kopal, professor of astronomy at Manchester University from 1951 to 1981, put it bluntly: 'If the cosmic telephone rings, for God's sake let us not answer, but rather make ourselves as inconspicuous as possible to avoid attracting attention!'

In 1976 Martin Ryle, Britain's Astronomer Royal from 1972 to 1982 and sharer of the 1974 Nobel Prize for physics, attracted worldwide media attention when he suggested that we might become the aliens' lunch. In a letter to the International Astronomical Union, he urged astronomers not to communicate with extraterrestrial civilisations. It was cosmic folly to reveal our existence and location to them, he warned. For all we know, he said, 'any creatures out there are malevolent or hungry', and once they knew of us 'they might come to attack or eat us'.

Frank Drake replied to Ryle's warning in the *New York Times* (22 November 1976): 'Put another way, Sir Martin Ryle has raised the question: Should humankind hide in this obscure corner of the tiny solar system attached to a minor star? ... The universe seems too rich to require an advanced race to look hungrily on Earth's meagre patrimony.'

Our planet is a long, long way for the little green men. Even if they know that we're here, interstellar distances make it difficult for us ever to meet an alien. However, replying to a phone call is good manners. Hello ET!

What should the reply say?: Douglas Vakoch, a resident psychologist at the SETI Institute, is not waiting for a message. He's already working on a reply. His job is to design messages that are intelligible to extraterrestrials. If we ever get a message from them, what should be the focus of our reply? Altruism, replies Vakoch. Evolutionary biologists consider altruism – the selfless concern for the well-being of others – central to group survival; and they believe it may be widespread among extraterrestrials, as they would also have developed it as a survival technique. 'If they're right,' says Vakoch, 'altruism would be a good starting point for interstellar messages describing central concepts of terrestrial biology, behaviour, and morality.'

It's not easy to convey the complex notions of altruism without any ambiguities to someone who doesn't speak our language. But Vakoch is confident that reciprocal altruism can be described through humans interacting. 'As an example,' he says, 'three-dimensional animation sequences could serve as "modern-day morality plays".'

'Perhaps the best clues that we are attempting to communicate altruism would come from the act of transmitting itself,' he adds. 'With no direct payoffs for senders, transmitting a celestial "message in a bottle" would convey something about our wishes to benefit extraterrestrials as well as future generations of humans, who might some day receive a reply of their own.'

Vakoch's message won't be limited to altruism. He is also working on a way to communicate the broad moral, ethical and religious principles to extraterrestrials. 'When people were deciding what to include on the Voyager spacecraft, they deliberately decided to exclude religious material,' he says. He wants to make sure that our next message tells the aliens how and what humans worship.

What happens after we discover microbial life on another planet?

Most scientists believe that if life is discovered on a solar system planet or moon, it will be microscopic – perhaps on the surface of Mars or in a sea beneath the icy crust of Jupiter's moon Europa. European space scientists Paul Clancy, André Brack and Gerda Horneck say that such a discovery could fit three scenarios.

First, the life form is very similar to terrestrial life. Second, the life form is different from terrestrial life, but it is still based on a DNA/RNA/protein scheme. Third, a life form that is radically different from terrestrial life. This scenario presents the greatest challenge to scientists. This life form could still be carbon-based but might depend on geothermal energy emitted from a planet's interior instead of solar energy via photosynthesis. Or it could be silicon-based, or even with a completely different and unimaginable chemistry.

The trio says that the discovered life object, which may be a fossil or a living micro-organism, will not be the endpoint of the search: 'Rather it will be the starting point of a whole new era of scientific research.' They say that scientists would make their studies in a number of contexts. Scientists would like to know the *geographical location* – for example, if it's on Mars, whether in or out of a crater, in a mountainous or plain region – and the *geological context* – such as the type of rock – to plan new searches. The next stage of their research, the *biological context*, will focus on the study of the biology of the object, such as the structure of its cells, its metabolism and similarities with terrestrial organisms. The *geochemical context* will relate to the chemistry and mineralogy of the sample in which the object is found.

Studies in the *water context* will try to find the answer to the critical question: What is the extent to which liquid water, either past or present, has been associated with the life-markers that have been discovered? The answer will indicate the importance of water for the genesis and survival of life. 'Has the case been overstated or not?' they ask. They consider the *contamination context* the most important aspect of the scientific study of the discovered life object; and emphasise that strenuous efforts will need to be undertaken to absolutely ensure that the object is not contaminated by humans or any robotic missions.

If the life object is a living micro-organism, Clancy, Brack and Horneck say that 'this would have impact not only on the daily news-chatter and preoccupations of Earth-bound humanity but also on the halls of serious science and philosophical communities, institutions and debate'. They claim that the scientific and technology communities are probably evenly split, but still a good 50 per cent tend to be hostile to the idea of life beyond Earth: 'So the discovery would settle all that and cause a real sensation.'

How would we react to the discovery of intelligent alien life?

In 2011, Robert A. Harrison, an American psychologist, conducted a survey that revealed that any discovery of extraterrestrial intelligence is likely to produce a mix of emotions including fear, pandemonium, equanimity and delight. However, people conditioned by years of participation in UFO clubs, science fiction and an endless discovery and endless parade of purported documentaries may find the discovery anti-climactic. Perhaps we should not worry too much about people who protect their belief system by denying scientific findings (or recasting them as theory), he says.

More recently, in 2018 Michael Varnum, a psychology professor at Arizona State University, and his colleagues conducted several experiments to try to gauge how would people react to the news of microbial life elsewhere in the universe. In *Frontiers of Psychology* journal, the researchers say that people felt more positively about microbial life outside Earth than they did about life created by humans in a lab. They didn't find any variation in responses based on the personality traits, political beliefs, income and other demographic factors.

The good news is that both psychologists assure us that there won't be a repeat of mass hysteria when H. G. Wells' novel, *The War of the World*, was dramatised on the radio in 1938 (see chapter 2)

Could intelligent aliens be as small as nanobots?

Nano-robots or nanobots are hypothetical intelligent machines too small to be seen by the naked eye. They were envisioned by Eric Drexler in his 1986 book, *The Engines of Creation*. Drexler, who is sometimes called the father of nanotechnology, imagined these tiny robots as self-replicating. Like biological cells, they would be able to make copies of themselves. In theory, they could build anything as long as they had a ready supply of the right kinds of atoms, a set of instructions and a source of energy.

Richard Smalley, who won the 1996 Nobel Prize in chemistry, dismisses the idea, saying that 'self-replicating, mechanical nanobots are simply not possible in our world'. To achieve their aim, nanotechnologists must provide these nanobots with 'magic fingers'. Within the constraints of a space of one nanometre – the size of nanobot – manipulation of atoms is not easy, because the fingers of a manipulator arm must themselves be made out of atoms. 'There just isn't enough room in the nanometre-size reaction region to accommodate all the fingers necessary to have complete control of the chemistry,' he says.

Let's apply this 'futurist's dream', as Smalley calls it, to intelligent aliens – aliens capable of building a technological civilisation. Could they be so small that they are almost invisible? By Smalley's logic the answer is a definite no. They won't be able to build anything. However, on a large planet where the pull of gravity is strong, it is possible that any life form may not be very tall. Therefore, aliens of the size of a large insect are possible. They could possibly have highly intelligent brains. With their 'magic fingers' they could build small robots that could build still larger robots and so on, and thus could have a technological civilisation far more advanced than our own.

Are alien skies blue?

Why is the sky dark at night? This deceptively simple question has puzzled astronomers for centuries. And no, the answer is not 'because at night the Sun is on the other side of the Earth'. In 1823 Heinrich Olbers, a German astronomer, pointed out that if there were an infinite number of stars evenly distributed in space, the night sky should be uniformly bright – with a surface brightness like the Sun's. He believed that the night sky's darkness was due to the absorption of light by interstellar space. But he was wrong. Olbers' question remained a paradox until 1929 when astronomers discovered that the galaxies were moving away from us, and the universe was expanding. The distant galaxies are moving away from us at a speed so high that it diminishes the intensity of light we receive from them. In addition, this light shifts slightly towards the red end of the spectrum. Red light has less energy than blue light. These two effects significantly reduce the light we receive from distant galaxies, leaving only the nearby stars, which we see as points of light in a darkened sky.

Why is the sky blue? This can be explained by the scattering of light by molecules in the atmosphere. This effect, known as Rayleigh's scattering, scatters shorter wavelengths best. Therefore, blue light (wavelength about 475 nanometres) scatters more efficiently than red light (about 650 nanometres). There is another reason. Our eyes are sensitive to electromagnetic radiation in a very short range. This 'visible light' range is from 400 to 700 nanometres (violet through red). Even within this range, we don't perceive all colours equally. If our eyes were equally sensitive to all colours, we would see the sky as violet (400 nanometres), as this colour is scattered more than the blue. But our eyes are more sensitive to blue and green than violet, so to us the sky is blue. We can only speculate how alien life forms, or even other life forms on Earth, would perceive the terrestrial sky.

The Moon and Mercury have no atmosphere; their skies are always black. On Mars, Rayleigh's scattering is very small, and during the day Mars' sky is yellow brown. Around sunrise and sunset, the sky is pinkish-red. The sky on Venus is yellow orange. For other planets we don't have any space-probe data from within their atmospheres: Saturn and Neptune's skies are probably blue, and Uranus' sky is probably greenish blue.

On exoplanets, the night sky would look very different to someone from Earth. The Sun would be visible to the naked human eye up to a distance of about 80 light-years.

Dying red giant stars may have habitable exoplanets. How would ET see their sky from such a planet? Philip Platt, an astronomer at Sonoma State University in California, imagines the view of a planet around a red star in a globular cluster (a smaller cousin of galaxies, a globular cluster contains from tens of thousands to a million stars in a sphere about 100 light-years in diameter): 'If the planet is earthlike, the sky won't be blue, because the star won't produce enough blue light. Instead, the sky might be orange or faintly red even at noon. The sky would quickly become black after sunset, allowing the glory of the cluster to emerge … Throughout the sky would be hundreds or perhaps thousands of brilliant stars, flaming jewels … They would cast a visible shadow and be seen easily during the daytime … Most would be ruby red, punctuated by blue sapphires.' He acknowledges that such a planet may exist only in our imagination.

What would aliens really look like?

We have made fictional aliens in our own image, but real aliens won't look like us. Species evolve, they are not created instantly. Humans are one of the billions of species that have arisen on Earth in 3.8 billion years of evolution. There is no straight line from the origin of life to intelligent humans. There were many evolutionary branches before one pathway led to the evolution of humans. It's statistically impossible that the same maze of evolutionary pathways would be repeated if evolution on our planet was run through again.

A habitable alien planet could differ from ours in many ways – distance from its parent star, gravity, moons, chemical composition, and so on – which could force the development of radically different life forms. All this makes it absolutely impossible that one day we will say hello to a loveable, wrinkly, long-fingered ET. 'Finding another planet with our kind of dinosaurs or people is more unlikely than finding a remote Pacific Island on which the natives speak perfect German or cockney rhyming slang,' says Jack Cohen, a reproductive biologist and the world's foremost authority on the physiology and behaviour of aliens.

Ulrich Walter of the Institute of Astronautics in Munich, who has made a scientific study of alien physiology, believes that the aliens are likely to have some similarities to life on our planet: they will be carbon-based life, basically composed of carbon, hydrogen, nitrogen and oxygen. As such life forms require water for their survival, their bodies need to contain water, making them soft like ours. Their bodies would also have 'blood', a liquid to transport nutrients to all the body cells. He estimates that their size could be between 10 centimetres and 10 metres, but most likely it would be about 1 metre. They would have a skeleton to bear the body, and probably also eyes, a lung and a mouth. 'The only question remains is how these parts are to be located,' he says. 'And this is something, the only thing, we leave to the imagination of science fiction writers and fantasy artists.'

Frank Drake believes that intelligent aliens 'won't be too much different from us … if you saw them from a distance of a hundred yards in the twilight you might think they were human.' He says that there will be some similarities: an upright stance, a head with eyes up top for good viewing and limbs free to manipulate tools, a requisite too for a technological society. 'My fun speculation – it's serious, but it sounds like a joke – is that I think, more often than not, the species will have four arms instead of two, because we all know that two arms are not good enough when you have to carry groceries,' he says. Perhaps Drake has been influenced by the images of four-armed Hindu goddess Kali.

Cohen does not support this tongue-in-cheek view of aliens. 'The real aliens will be very alien indeed,' he says. He is almost certain that they will not be humanoid. He considers the development of spines and skeletons as 'an evolutionary accident that could well be unique to Earth'. 'We can make an educated guess that we'll never see anything like a human being on an Earth-like planet.' They will not have faces like ours at all – eyes, yes, but nose, external ears, teeth-like-scales, no – and certainly will not share our interest in prurient experiences,' he says. 'No alien pornography of our kind, then.'

He says there is a tendency for people to think that biochemical life is the only kind of life, but he would not be surprised if we find other kinds of non-water-based life. 'You can find self-replicating vortexes in the atmosphere of the Sun. There are all sorts of self-reproducing systems in all sorts of environments which could hold potential for life.'

Arik Kershenbaum, a University of Cambridge zoologist, presents a scientific view of what alien will look like in his 2020 book, *The Zoologist's Guide to the Galaxy: What Animals on Earth Reveal About Aliens – and Ourselves*. He predicts that alien life will bear striking similarities parallel to life on Earth. Most aliens will be bilaterally symmetrical and use familiar forms of locomotion such as legs, paddles or jets.

Until we meet an alien, everyone is free to design their own. You may do well if you heed Cohen's advice: 'This involves deeper, more divergent science than DNA-is-God-and-RNA-is-its Prophet molecular biology and is much more fun, too.'

What would alien plants look like?

Nancy Y. Kiang, an American biometeorologist, is an extraordinary expert when it comes to the colour of alien plants. On our planet, visible light (violet through red) and near-infrared light (infrared light that comes after red light on the electromagnetic spectrum) power photosynthesis. This wide range of electromagnetic spectrum gives planets on Earth a wide range of colours. However, the signature of chlorophyll, green pigment, is unique to our plants. On planets around stars hotter than the Sun, plants would absorb blue light and would look green to yellow to red to our eyes. On planets around dim stars such as red dwarfs, the plants will receive less visible light and the plants would look black.

Our understanding of the unique signature of vegetation for other planets will be key to designing future space missions and interpreting their data. 'Our very ability to search for life elsewhere in the universe ultimately requires our deepest understanding of life here on Earth,' Kiang says.

Would they have religion?

OK, ETs won't look like us, but would they have religions that resemble ours? No, says SETI researcher Seth Shostak; however, he is more interested in the larger question of whether the extraterrestrials will have religion at all. 'Virtually every group of peoples on Earth has some sort of religion, and that suggests that religion is a "universal" phenomenon.' But he is worried that if we're hard-wired for religion as 'an accident of evolution', aliens might find our interest in religion a peculiar anomaly. 'This is a disturbing scenario, although not really a threatening one.'

At least Jill Tarter, doyen of SETI research, is not threatened by this scenario. In a thoughtful little essay, 'O give ye praise Europans', she says that organised religion is an invention of the mind, as envisioned by evolutionary biologist Steven Pinker. A technological civilisation cannot live for long, she warns, unless it has 'a universal religion with no deviations, no differentiations – absolutely global and compelling for all'. Such a religion, she believes, might be able to co-exist for a long time with technological development without precipitating the worst of human tendencies. 'If that development is possible for any civilization, then I would speculate that, if and when we ever get a message, it's going to be a missionary appeal to try to convert us all. And, on the other hand, if we get a message and it's secular in nature, I think that says that they have no organized religion – that they have outgrown it.'

Unlike Asimov's three famous laws of robotics, Arthur C. Clarke's three laws, postulated in 1962, deal with humans:

Clarke's First Law: When a distinguished but elderly scientist states that something is possible, he is almost certainly right. When he says that something is impossible, he is very probably wrong.

Clarke's Second Law: The only way of discovering the limits of the possible is to venture a little way past them into the impossible.

Clarke's Third Law: Any sufficiently advanced technology is indistinguishable from magic.

The Third Law has prompted Michael Shermer, founding publisher of *Skeptic* magazine, to propose Shermer's Law: Any sufficiently advanced extraterrestrial intelligence is indistinguishable from God.

In an essay, 'Is God more than a sufficiently advanced extraterrestrial intelligence?' (www.edge.org), Shermer says that if we ever do find extraterrestrial intelligence (ETI) 'it will be as if a million-year-old "Homo Erectus" were dropped into the middle of Manhattan, given a computer and cell phone and instructed to communicate with us. ETI would be to us as we would be to this early hominid – godlike.'

Freeman Dyson has also expressed similar views: 'Whoever they are, they probably know a great many things that we don't know. Of course, the danger is that if they know it all, then we'd lose self-confidence.'

Although science has not even remotely destroyed religion, Shermer says, his law predicts that the relationship between the two will be profoundly affected by contact with technologically advanced extraterrestrials. 'To find out how, we must follow Clarke's Second Law, venturing courageously past the limits of the possible and into the unknown,' he suggests.

In 2014, an article *in The Atlantic* magazine claimed that Pope Francis would absolutely baptise aliens if they showed up at Vatican and asked for it. Aliens reading this news should be heartened that on Earth at least they can become good Catholics and behave benevolently towards earthlings. They may change their mind once they learn about the inquisition of Galileo and other astronomers who claimed that there were worlds other than ours. Maybe not as that was in the distant past.

What does the law say about aliens?

At present, extraterrestrial life forms, primitive or intelligent, have no rights under any national or international law.

In 1969 the United States federal authorities passed a law that made contact between a US citizen and an extraterrestrial or their vehicles strictly illegal. The penalty for this 'extraterrestrial exposure' was maximum imprisonment of one year or a maximum fine of $5,000, or both. Curiously, the law was enacted on 16 July 1969, the day Apollo 11, the first crewed mission to the Moon, was launched. Were the Feds worried about aliens hitch-hiking to Earth with Neil Armstrong, Michael Collins and Edwin 'Buzz' Aldrin? The law was repealed in 1991.

Whether they hitch-hike on earthlings' space missions or come by their own flying saucers, extraterrestrials have no protection under any local law. One can only hope that they have enough intelligence to buy interplanetary travel insurance.

Contact with an alien civilisation will have profound religious, cultural and other impacts on humankind. There are no international agreements specifically addressing these issues.

The 1967 United Nations Outer Space Treaty has limited scope. It states that the exploration and use of outer space, including the Moon and other celestial bodies, shall be carried out for the benefit and the interests of all countries … and shall be the province of all humankind. The Treaty also says that that all human activities relating to outer space are of international concern, and expressly mandates freedom of scientific investigation and international cooperation in out space investigation and activities.

Article 9 of the Treaty says that space explorations shall avoid 'harmful contamination' of Earth and other planets. All NASA missions are required to follow planetary protection procedures to prevent transport of Earth microbes to a planet, moon or asteroid. The procedures include careful cleaning and sometimes heat sterilisation of spacecraft. Similarly, precautions are taken against alien microbes hitch-hiking on space missions returning to Earth.

The Declaration of Principles for Activities Following the Detection of Extraterrestrial Intelligence, developed in 1989 and revised in 2010 by the SETI Permanent Committee of the International Academy of Aeronautics, has been approved by many international bodies such as the International Academy of Astronautics, the International Astronautical Federation, the International Institute of Space Law and the International Astronomical Association. In brief, the principles are:

- All SETI experiments should be conducted transparently.
- Discoverers should make all efforts to verify the authenticity of a supposed detection of ET intelligence.
- If a detection's authenticity is confirmed, discoverers should openly disclose the detection to the public, the scientific community and the UN Security Council.
- Scientific data should be made available to the scientific community.
- If credible evidence of extraterrestrial intelligence is established, a committee of experts should be formed to analyse all data collected in the

aftermath of the discovery, and also to provide advice on the release of information to the public.

- No reply should be made until international consultation has taken place.

Legal experts say that the Declaration is not binding in law. If signatories decide not to abide by it, there is no remedy by way of law or any enforcement mechanism other than the disciplinary mechanism of scientific institutions.

CHAPTER 14

ET IN OUR HISTORY

A certain grandeur

In our times, science's unlikely poster boy is a wild-haired old man with piercing eyes. In times tagged in our history books as BC (or BCE – Before Common Era), Einstein's equivalent was Aristotle.

Even half a century after his death, Einstein's face – with its unique mixture of intensity and distraction – is familiar to billions of people around the world. In days when there was no TV, no social media, not even newspapers, Aristotle's face never became ubiquitous. However, his influence was extraordinary: his ideas dominated scientific thought for nearly 2,000 years. A feat yet to be surpassed by Einstein!

But what did he look like? Ancient portrait busts show a stern, sage-like figure: a little bald, a deep brow, small eyes, an aquiline nose and a thick beard. Writing five centuries after Aristotle's death, Diogenes Laertius, a 2nd-century AD Greek biographer, remarked: 'He had a lisping voice … he wore fashionable clothes, rings on his fingers, and used to dress his hair carefully.' Very much unlike Einstein.

When Aristotle was born in 384 BC in Stagira, a small town in northern Greece, the Greek world was more than Greece – it extended along the coasts of Turkey, Egypt, Libya, Spain, France and Italy – and was composed of self-governing territories with a single urban centre (the *polis*, the city-state, from which comes 'politics'). The largest *polis* was Athens, with a population of about 300,000.

At the age of 17, Aristotle travelled to Athens, where he enrolled in the Academy of the renowned philosopher Plato. The Academy was not like a college or university of today: there were no formal courses, no examinations, no degrees. Instead, there were discussions, mainly in the form of a dialogue between two people. In dialogue, one person asks a question, the other expands upon the idea, presents possible answers, asks further questions, and so on. This unique method of furthering knowledge was pioneered by Plato's teacher, the philosopher Socrates, who was accused of corrupting the minds of young Athenians and was sentenced to death by drinking poison.

Young Aristotle blossomed in the Academy's stimulating intellectual environment; Plato nicknamed him 'the reader' and 'the mind of the Academy'. A keen reader he was, indeed. Diogenes confirms this: 'He always placed beside his bed a brass dish, and when he lay down to read and rest, he would hold in his hand extended over the dish a brass ball. When he was overcome by sleep, the ball would fall into the dish, and the noise would immediately awaken him.' The use of this Aristotelian contraption is not compulsory while reading this book.

Aristotle stayed at the Academy for about 20 years, first as a student and then as a teacher. In 343 BC, King Philip of Macedonia, who was Aristotle's boyhood friend, asked him to tutor his 13-year-old son, Alexander (the future Alexander the Great). Aristotle taught the young prince for three years and in 334 BC, returned to Athens to set up his academy at the Lyceum, a temple of the Sun-god Apollo. Since Aristotle had the habit of walking with his students up and down the academy's main avenue while teaching and discussing, the academy became known as the Peripatetic School (from the Greek *peripatetikos*, 'walking up and down').

After Alexander died in 323 BC in Babylon, anti-Macedonian sentiments in Athens began to rise. Aristotle, who came from the area, was charged with impiety, and the lack of reverence for religion. Rather than suffer the fate of Socrates, he fled to Chalcis, his mother's hometown, saying that he would not allow Athens to sin twice against philosophy. He died the following year, 322 BC, leaving an astonishing legacy for posterity: a corpus that covered most branches of knowledge. Much of this greatest 'encyclopaedia' of the ancient world represented Aristotle's original thoughts and observations.

In his biography, Diogenes lists some 550 books by Aristotle. Of these, only about a third has survived. By our standards, these are rather slim volumes and would make about 2,500 modern pages. Much of this work deals with logic, ethics, politics, metaphysics (the ideas of existence, truth and knowledge), biology, physics and astronomy.

Aristotle's writings on logic are called *Organon* ('the instrument'). They set out his ideas of science as a logical system in which the truths of nature can be deducted from universal principles. These universal principles were perceived intuitively as correct, but, like axioms of geometry such as 'a straight line can be drawn between two points', were unprovable. For example, from the universal principle that the circle was the perfect shape in nature, Aristotle deduced that the heavenly bodies must move in circles. Such was Aristotle's influence that when in 1609 Kepler calculated that the planets moved in elliptical orbits, he had difficulty convincing himself that this was true.

Since the sphere was the perfect shape for a body, Plato advocated that Earth must be a sphere. Aristotle presented several scientific and decisive arguments, including that during an eclipse of the Moon, the edge of Earth's shadow was always circular, and this could not happen unless Earth were spherical. He said that it was at rest at the centre of the universe, with the Moon, the Sun and the planets revolving around it in concentric spheres. These spheres were bounded by a celestial sphere, on the concave side of which were the fixed stars. This sphere revolved on its axis once in 24 hours. The stars were spherical and did not have any independent motion. They were not hot; their light and heat came from friction with air. Meteors and comets were not celestial objects because they did not have circular orbits like planets; they were related to meteorological phenomena like lightning and rainbows.

He suggested that everything on Earth was made from four elements – earth, water, air and fire. Each of these elements had its own type of movement in a certain direction: earth went down, the fire went up, water went above the earth, and the air went above water but below the fire. Like the Moon, the Sun and all the other planets and stars travelled in a circle, different from the straight-line motion of the four elements; he proposed that they were made of a fifth element, ether, whose natural movement was circular. The celestial spheres were moved by a divine force he called the prime mover (in medieval illustrations of the cosmos, the prime mover was transformed into angels).

There are other worlds

In the 5th century BC Leucippus and his pupil Democritus, the Greek philosophers famous for their atomic theory – that matter is made of empty space and an infinite number of tiny particles called *atomos* or atoms which are always in motion – also suggested that atoms produced innumerable worlds of different sizes. 'In some worlds, there is no sun or moon, in others, they are larger than ours, and others have more than one,' Democritus said. His comment that 'some of the worlds have no animals and plants and no water' is the first-ever scientific allusion to life beyond Earth.

A century later, Epicurus advanced the atomic theory of Democritus and combined it with the pleasure ethics of Aristippus, a pupil of Socrates who believed that sensory enjoyment was the most important thing in life. His followers were called Epicureans, and they lived in a garden. The story goes that an inscription on the gate to the garden said: 'Stranger, here you will do well to tarry; here our highest good is pleasure.'

Diogenes records a letter of Epicurus to Greek historian Herodotus in which he wrote: 'There are infinite worlds both like and unlike this world of ours. As was already proved, the atoms being infinite in number are borne on far out into space. For those atoms of such nature that a world could be created by them or made by them, have not been used up either on one world or a limited number of worlds … So that there is no obstacle to the infinite number of worlds.' Later in the letter, he suggests that these worlds have 'living creatures and plants and other things we see in this world'.

Metrodorus of Chios, a pupil of Epicurus, supported his teacher: 'To consider Earth the only populated world in infinite space is as absurd as to assert that in an entire field sown with millet, only one grain will grow.' His claim is that the worlds are infinite in number, followed from the atoms being infinite.

In the 1st century BC Lucretius, a Roman poet who was a contemporary of Julius Caesar, wrote an epic poem, *De rerum natura* (On the Nature of the Universe), which kept alive Epicurus' atomism. The poem's central theme is that behind all-natural phenomena lie eternal, unchanging atoms that can arrange and rearrange themselves into different forms. He explains how atoms form solids, liquids and vapours. Solid substances, for example, are 'held together linked and interwoven as though by rings and hooks'.

An interesting aspect of such combinations of atoms is that, as there is an abundance of atoms available, these atoms are capable of forming other worlds elsewhere in the universe, with races of different humans and animals:

Now if there is so vast a store of seeds

That the whole lifetime of all conscious beings

Would fail to count them, and if likewise Nature

Abides the same, and so has power to throw

The seeds of things together everywhere,

In the same manner as they were thrown together

Into our world, then you must needs admit

That in other regions there are other earths,

And diverse stocks of men and kinds of beasts.

– Lucretius, *De rerum natura*, trans. R. C. Trevelyn, Cambridge University Press, Cambridge, 1937

Aristotle's elaborate ideas redefined the cosmos, but they remained finite and changeless: he did not entertain the idea of never-ending sequences of celestial spheres. Nor did he entertain the idea of the existence of innumerable worlds advanced by Democritus and Epicurus. In his cosmos, the natural movement of the element earth was towards Earth's centre, which was also the centre of the world. Still, the element fire moved away from the centre of Earth and towards the outer circumference. The elements air and water assumed their natural places between the centre of the Earth and its outer circumference. If there were more than one world, the elements would have more than one natural place towards which to move. Many worlds suggested the idea of more than one centre and one circumference. To him, it was a logical contradiction. 'There cannot be several worlds,' he declared. That ended the debate for centuries.

Aristotle is still admired as a great philosopher, but in matters of science, the poster boy of antiquity was wrong most of the time. Unlike Einstein ('It's the theory that decides what we can observe'), he observed without forming detailed theories. Preconceived notions distorted his power of observation. Some say that his ideas stalled the progress of science until they were challenged by Bacon, and proved wrong by Copernicus, Kepler, Galileo, Newton and others.

Francis Bacon, an English philosopher, advanced a new method of inquiry, completely different from the philosophical approaches of Aristotle, in his book *Novum Organum* ('the new instrument'), which has influenced every scientist since its publication in 1620. 'There are two methods of investigation, through argument and experiment,' he wrote. 'Argument does not suffice, but the experiment does.' He rejected Aristotle's deductive approach to reasoning and suggested an *inductive* approach. The most important aspect of this method was drawing up tentative hypotheses from available data and then verifying them through further investigations.

Although Bacon dethroned Aristotle, he did not challenge Aristotle's belief in one world. However, he invoked a different argument: the impossibility of a void between different worlds. 'If there were another universe, it would be of spherical figure,' he said. 'Therefore, they must touch, but they cannot touch each other except at one point … Hence elsewhere than in that point, there will be a vacant space between them.'

Bacon discovered the most important tool of science – the scientific method – but he did not make any significant scientific discoveries. 'I shall content myself to awake better spirits like a bell-ringer, which is first up to call others to church,' he once wrote to a friend. Bacon's bell is still ringing, yet it can never drown out the contribution of one of the greatest geniuses in the progress of science.

Aristotle said: 'The search for truth is in one way hard and in another way easy, for it is evident that no one can master it fully or miss it wholly. Each adds a little to our knowledge of nature, and from all the facts assembled there arises a certain grandeur.'

No other ancient institution has contributed more to that certain grandeur than the Great Library and Museum at Alexandria.

The poster girl of science

In 331 BC, Alexander stopped during one of his journeys at the western end of the Nile delta, a few kilometres inland from the Mediterranean, where the sea turned into the shoreline to form a natural harbour and decided to build a city there. He took a personal interest in planning it, and marked out various sites, including the Museum (a shrine to the Muses), with chalk. Legend has it that when he ran out of chalk, he used barley from the soldiers' mess as a substitute. But it was soon gobbled up by seagulls. Alexander considered it a bad omen but his personal seer, Aristander, assured him that the birds' coming was most auspicious. It was a sign that the new city would be great and prosperous. Aristander's prophesy was fulfilled: Alexandria became the cultural and commercial hub of the ancient world.

After Alexander died in 323 BC his empire broke into three regions which his generals ruled. One of them, Ptolemy Soter, declared himself king of Egypt in 304 BC. The Ptolemaic dynasty ruled Egypt, with its capital at Alexandria, until the death of Cleopatra in 30 BC.

Following Alexander's wishes, Ptolemy Soter founded the Museum. As he had considerable intellectual interest, he founded a library with the Museum. Demetrius, an Aristotelian philosopher, was in charge of collecting books for the new library. It was almost certainly modelled on Aristotle's personal library. Not for nothing was Aristotle called 'the reader'. His outstanding collection of books at the Lyceum was arguably the first university library in history. Strabo, a 1st-century AD Greek historian, claims that Aristotle was 'the first man whom I know to have collected books and taught the kings of Egypt how to arrange a library'.

Over the centuries, the Library amassed more than 400,000 scrolls, but none has survived. This great loss has been attributed, in turn, to the Romans under Julius Caesar in 47 BC (40,000 volumes were burned in a fire lit by soldiers to clear the wharves to block the fleets of Cleopatra's brother), Christian zealots in AD 391 (an unknown number of works were lost during their riot), and the Arab armies of the Caliph of Baghdad in AD 641 (they not only razed the Library buildings but burned the books to heat the public baths).

The Library also supported a *synodos* (community) of scholars who made it the chief centre of science and scholarship in the ancient world. Many of the scholars enhanced Aristotle's image of the cosmos.

Aristarchus was a contemporary of Archimedes (who gave us the first 'Eureka!' moment in science), and both were associated with the Library. Archimedes credits Aristarchus as having taught that the Earth was not at the centre of the universe but that it moved around the Sun. His idea of a moving Earth looked utterly strange during those ancient days, and it was, of course, rejected by his contemporaries.

Eratosthenes, who was appointed head of the Library in 236 BC, calculated the size of Earth. His value for the diameter, in modern units, was 12,633 kilometres, which is only 101 kilometres short of the true average diameter – a remarkably accurate measurement. He was a versatile scholar: an astronomer, mathematician, geographer, historian, literary critic and poet. He was nicknamed 'Beta' (the second letter of the Greek alphabet) because he was considered the second best at everything.

In the 2nd century BC, Hipparchus applied a yardstick to the heavens. He used sundials, water clocks and a circular instrument divided into degrees and fitted with a simple sighting device for a rigorous study of the sky. He drew up tables of the motions of the Sun, the Moon and the planets, and from these tables, he predicted eclipses for long periods ahead.

'Earth does not rotate; otherwise, objects will fling off its surface like mud from a spinning wheel. It remains at the centre of things because this is its natural place

– it does not tend to go either one way or another. Around it and in successively larger spheres revolve the moon, Mercury, Venus, the sun, Mars, Jupiter and Saturn, deriving their motion from the immense and outermost spheres of fixed stars.' So proclaimed Claudius Ptolemy (not related to the Ptolemys) in c. AD 150 in his book *Almagest*, in which he synthesised the work of his predecessors. A major part of *Almagest* ('the greatest' in Arabic) deals with the mathematics of planetary motion. Ptolemy explained the wanderings of the planets by a complicated system of cycles and epicycles, which harassed astronomers for centuries. 'If the Lord Almighty had consulted me before embarking upon the Creation, I should have recommended something simpler,' commented Alfonso the Wise, the 13th-century Spanish king of Castile, a great patron of astronomy. However, Ptolemy's erroneous theory dominated astronomy for 14 centuries until Copernicus challenged it and demolished by Kepler.

In AD 400, Hypatia, daughter of Theon, a mathematician and astronomer at the Library, became head of the neo-Platonist school of philosophy. She was also an outstanding mathematician and inventor. Only the titles of her mathematical works have survived, but sources describe her as a mathematician who surpassed her father's talents. She invented, among other things, the plane astrolabe to measure the position of stars, planets and the Sun.

Hypatia was a pagan in an increasingly Christian city. One dark night in 415, on the way home from the Library, she was dragged off her chariot by a mob of extremist Christians, stripped stark naked, hacked to death, and her remains burned.

In the 20th century, the life and death of beautiful and brilliant Hypatia were romanticised by feminists, and she became the poster girl of modern science.

With Hypatia's death ended, the golden age of Greek science. Europe had entered the Dark Ages. Earth – the only world – stood at the centre of a bounded and finite universe. The Christian Church wholeheartedly accepted this Aristotelian view. The first challenge to it came in the 14th century.

Ockham's pen

William of Ockham (also spelt Occam), a philosopher and theologian, came from Ockham, a village in Surrey, near London. In his youth he joined the Franciscan order and studied at Oxford, where he lectured from 1315 to 1319. At Oxford, which was then a great Franciscan centre of learning, William became the leader of a school of philosophy called nominalism.

Aristotle's main argument for a single world was based on the causes of the motions of the four elements. When placed outside their natural place, each element returned to its unique natural place. William, however, suggested that the elements would not necessarily return to their unique natural places, but to a natural place dependent on their situation. He rejected Aristotle's view of a finite universe and said that there could be no assurance that the world was finite, or that it had a governing unity, or that it was eternal, or that there were not several worlds. He said that the elements in each world would return to their natural place within their own world.

William is now most remembered for his rule, known as Ockham's Razor, which is of vital importance in the philosophy of science. The rule – *a plurality (of reasons) should not be posited without necessity, or it is vain to do with more what can be done with less* – implies that the number of causes or explanations needed to account for the behaviour of a phenomenon should be kept to a minimum. It is a guiding principle in developing scientific ideas, and it insists that you should prefer the simplest explanation to fit the facts. The rule has been interpreted in modern times to mean that when you have two competing theories that make exactly the same predictions, the one that is simpler is the better. In other words, the explanation requiring the fewest assumptions is most likely to be correct. Advice to computer programmers to keep their programs simple – *keep it simple, stupid* (KISS) – is in a similar vein. But we must also heed Einstein: 'Everything should be made as simple as possible, but not simpler.'

William's statements in his philosophical and theological writings, including the ideas on other worlds, aroused such opposition that he was refused his Master of Theology degree at Oxford and was ordered to appear before the papal court on charges of heresy. He fled to Germany and, according to a story, probably apocryphal, asked Emperor Louis IV for protection with the plea: 'Protect me with your sword, O Emperor, and I shall protect you with my pen.' He remained in Germany for the rest of his life.

Two centuries later, a pen mightier than that of William redrew Aristotle's cosmos and demoted Earth from its pride of place.

Innumerable worlds

'The sun is at the centre of the solar system, fixed and immobile, and planets orbit around it in perfect circles in the following order: Mercury, Venus, Earth with its moon, Mars, Jupiter and Saturn,' declared Nicolaus Copernicus, a Polish astronomer and cleric. The Copernican system defied the dogma that Earth stood still at the centre of the universe and set forth a new theory of a Sun-centred universe.

Not only did Copernicus place the Sun at the centre of the solar system, but he also gave detailed accounts of the motions of Earth, the Moon and the planets that were known at the time. He said that Earth also revolves on its own axis, which accounts for days and nights.

Copernicus had found the truth, but to convince the world was an onerous task. For more than three years he taught his new theory to his students at the University of Rome, but he could not stagger their belief in an Earth-centred universe.

He did not publish his findings because they were thought to contravene the teachings of the Church. Religious leaders of his time were against him. Martin Luther (founder of the Lutheran Church in Germany) denounced him as 'a new astrologer ... the fool' who wanted 'to overturn the entire science of astronomy'. John Calvin quoted Psalm 93 against him: 'Surely the world is established, so that it cannot be moved.'

Copernicus passed his life in agony. His efforts to convince the Church of the truth were in vain. His book *De revolutionibus orbium coelestium* (On the Revolutions of the Heavenly Spheres) was published in Germany at the very end of his life. Andreas Osiander, a Lutheran theologian and one of Copernicus' students, who was supervising the printing process, was so frightened of the revolutionary ideas contained in the book that he wrote a preface claiming that it was not a scientific treatise but a 'playful fancy'. The first copy of the book was sent to Copernicus in Poland. It arrived but a few hours before his death on 23 May 1543. Thus, the greatest astronomer of his time died without seeing his book in print – the book which ranks with Newton's *Principia* and Darwin's *Origin of Species* as a product of scientific genius.

Copernicus' book was in defiance of the Church's teachings, but it was too late for the Church to do anything. A few decades later Giordano Bruno, an astronomer and Dominican monk, was burned at the stake for his belief in Copernicus' heretical views.

9 February 1600. It is freezing cold in the vast and ornate Hall of Inquisition in Rome. Fifteen illustrious cardinals of the Holy Office are seated on high-backed plush chairs forming an arc around the accused – a 51-year-old, small, thin man with black hair and dark brown eyes, kneeling silently. The Grand Inquisitor, Cardinal Severina, reads the charges. The eight counts of heresy include belief in the movement of the Earth and in an infinite universe filled with innumerable worlds. Severina asks Bruno to recant his belief and pray for mercy to God. Bruno remains silent. Severina excommunicates the heretic and sentences him to die 'without shedding of blood' (in other words, to be burnt alive at the stake). A defiant Bruno lifts his head and declares: 'Perhaps you, my judges, pronounce this sentence against me with greater fear than I receive it. The time will come when all will see as I see.' He is given eight days' grace to recant and deny his beliefs. His belief in the truth remains unshakable.

19 February 1600. The man lies naked on a rack, his ankles and wrists bound tightly, in a dark, dreary and damp dungeon in the terrible Tor di Nona prison in Rome, which has been his home for the past two years. In the feeble light of winter dawn, his guards ask him to put on the sulphur-coloured garb of heresy, covered with pictures of devils and crimson flames and crosses. He is then led in chains through a howling, fanatical crowd to the site of the execution, a public square called the Campo de Fiori. He walks calmly and with dignity, his face serene, his head high. He is stripped naked, and a shirt of pitch that extends from his waist to his feet is put over him so that he would not die as quickly. The executioner ties him to the stake, piles firewood, charcoal, kindling up to the chin and places a torch between the feet. As the flames blaze around him, a priest pushes forward and presses a crucifix into his hands, but the man turns his head away. Within seconds, flames sear him, smoke and fire surround him. When the fire subsides, his remains are powdered and blown in the wind so that no relic of the heretic would survive.

But the man lives on.

Three centuries later, a statue in honour of Bruno was erected at the exact site where he was burned. The statue in Campo de' Fiori, within walking distance of the Vatican, is now surrounded by a colourful and busy fruit and vegetables market.

Bruno's most important work on the Copernican system is *La cena de le ceneri* (The Ash Wednesday Supper), published in 1584, in which he describes Copernicus as 'a grave spirit, meditative, penetrating and mature; a man who did not surrender himself to any past astronomers.' Bruno also did not surrender himself completely to Copernicus. In *De l'infinito universo e mondi* (On the Infinite Universe and the Worlds), published in the same year, he *diverges* from Copernicus. This book, which started his fatal dispute with the Church, is in the form of a dialogue between two philosophers named Burchio and Fracastorio. When Burchio asks, 'Then the other worlds are inhabited like our own?', Fracastorio replies: 'If not exactly as our own, and if not more nobly, at least no less inhabited and no less nobly. For it is impossible that a rational being fairly vigilant, can imagine that these innumerable worlds, manifest as like to our own or yet more magnificent, should be destitute of similar or even superior inhabitants.'

Copernicus' ideas are often referred to as the 'Copernican Revolution', but Milič Čapek, an American philosopher, calls this expression an historical inaccuracy because 'it gives Copernicus the credit which really belongs to Giordano Bruno, the first to depart wholeheartedly and consistently from Aristotelian cosmology'.

In *E mondi*, when Burchio asks, 'You maintain that Plato is an ignorant fellow, Aristotle an ass and their followers insensitive, stupid and fanatical?', Fracastorio replies: 'As I told you from the first, I regard them as earth's heroes. But I do not wish to believe them without cause, not to accept those propositions whose antitheses are so compellingly true.' 'Who then shall be judge?' asks Burchio. 'Every well-regulated mind and alert judgment', Fracastorio replies.

Vortices in the sky

Fracastorio would have agreed that René Descartes, the 17th-century French philosopher and mathematician who invented analytical geometry ('the method of giving algebraic equations to curves', as Voltaire described it), was indeed a 'well-regulated mind'. Unlike Bruno, he did not criticise Aristotelian physics, but created an entirely new physics to explain the motions of heavenly bodies. The laws of nature were the laws of mechanics, he asserted, and everything in nature could ultimately be reduced to the rearrangement of particles moving according to these laws. Space by itself being nothing, it has no extension – no length, breadth or height. Only matter has the property of extension, and space cannot exist where there is no matter. Matter exists everywhere, and a vacuum exists nowhere.

The most popular aspect of Cartesian physics was the vortex theory. It provided a simple, mechanistic explanation of the universe which was understandable to everyone. The motions of the heavenly bodies, the theory proposed, consist of whirlpools or vortices of a subtle matter – 'ether'. Each vortex has a star surrounded by a vast space. Every star is a sun, and every sun has its own family of planets. The universe is infinite and consists of a vast number of vortices, all limiting and circumscribing each other. In the solar system, the planets are carried about in the Sun's vortex and the Moon is carried around Earth in the same way. Descartes was aware of Kepler's laws that the planets move in elliptical orbits; however, his vortexes [vortices?] lead to circular orbits.

The following figure shows Descartes' system of vortices. Each vortex denotes a star surrounded by a vast space. In the case of the solar system, the vortex carries the planets around the Sun (S). The irregular path across the top of the diagram shows the motion of a comet, which passes through many vortices. (From Descartes' *Le monde, ou traité de la lumière*)

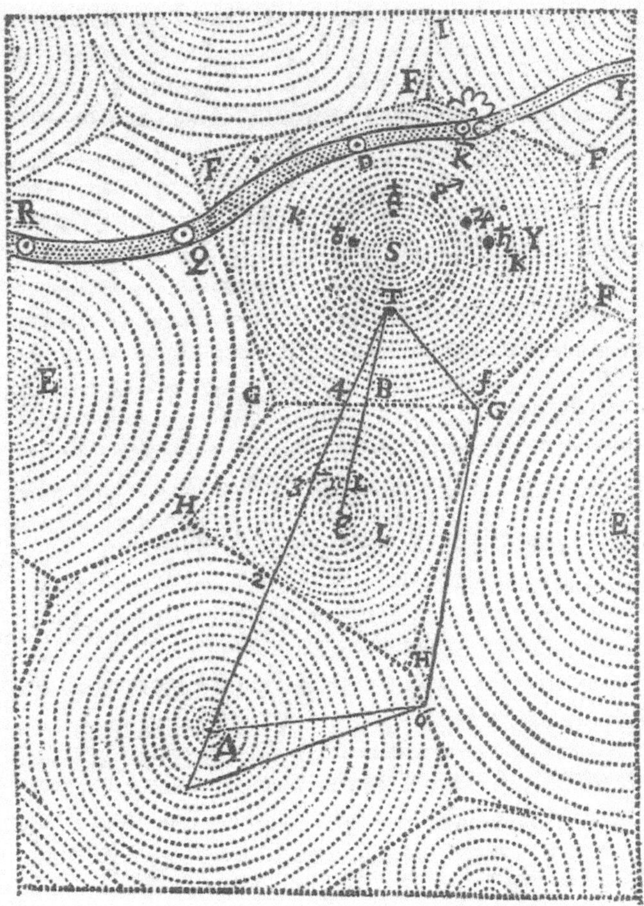

Descartes was 37 when he presented his vortex theory in an imposing treatise, *Le monde, ou traité de la lumière* (The World, or a Treatise on Light). When he was giving finishing touches to his book, he heard the stunning news that on 22 June 1633 Galileo had been forced by the Inquisition to abjure as a heresy the

Copernican doctrine ('I Galileo Galilei, son of Vincenzio Galilei, of Florence … will not hold, defend or teach the said false doctrine in any manner … I give assurance that I believe, and always will believe what the Church recognises and teaches as true.') and had been punished by confinement to his house.

The image of a frail old man of 70 kneeling before the awful tribunal and recanting his belief crushed Descartes. 'He was not only afraid – as any sane man might well have been; he was deeply hurt,' writes E. T. Bell in his authoritative biographies of mathematicians, *Men of Mathematics*. 'He was as convinced of the truth of the Copernican system as he was of his own existence. But he was also convinced of the infallibility of the Pope.' As he had no desire to become a martyr, he decided not to publish *Le monde*: 'Although I consider all my conclusions based on very certain and clear demonstrations, I would not for all the world sustain them against the authority of the Church.'

His masterpiece, *Discours de la méthode* (A Discourse on the Method), which gave analytical geometry to the world, was published in 1637. In 1644 he published his most comprehensive work on physics, *Principia philosophiae* (Principles of Philosophy), which was an extension of *Le monde*. The Church ignored its publication but placed it on its Index of prohibited books. Yet it was widely read and caused a stir in the world of science.

In this book Descartes presented the first complete physical system since that taught by Aristotle. The vortex theory caught the imagination of contemporary scientists and was taught widely in universities in Europe, but only until 1687 when Newton presented a revolutionary idea in his magnum opus, *Philosophiae naturalis principia mathematica* (Mathematical Principles of Natural Philosophy), that a single universal force, gravitation, keeps heavenly bodies in their orbits. The vortex theory was now a lost cause. Today it is a mere relic in the museum of dead theories.

The vortex theory obviously implied the existence of an infinite number of planets, yet Descartes remained silent on the issue. When his friend and patron, Christina, Queen of Sweden, was disturbed by the implication that a Cartesian must 'probably hold that all these stars are inhabited, or, still better, that they have earths around them, full of creatures more intelligent and better than he,' he replied guardedly: 'I always leave these questions, once posed, suspended, preferring not to deny and not to affirm anything.'

The trials of Bruno and Galileo were still fresh in the mind of the man who also gave us the single best-known philosophical statement, *Cogito ergo sum* – 'I think, therefore I am'. Nevertheless, for his followers the trials were as distant as his vortices in the sky. The Catholic Church's authority was waning, and the Cartesians were not afraid to assert explicitly the existence of an infinite number of planets, inhabited or otherwise. One of them was Bernard le Bovier de Fontenelle.

A charming philosopher and a sharp-witted countess

Fontenelle was a poet, playwright, essayist and natural philosopher. He was also an ardent Cartesian. As a writer and as the permanent secretary of the French Academy of Sciences – although he made no scientific discoveries, he was appointed to this prestigious post in 1697 – his influence in furthering the cause of vortex theory was great.

He was a charming and witty man. In his nineties, when he met a young woman, he exclaimed: 'Ah! If I were only 80 now!' He continued to write until the day he died – one month short of 100 years – on 9 January 1757. He died without suffering, telling his doctor at his deathbed that he felt nothing apart from 'a difficulty in continuing to exist' (*une difficulté de'être*).

Fontenelle's charm and wit – and his admiration for the Copernican system and Descartes' vortices – is most abundant in a lively little book, *Entretiens sur la pluralité des mondes* (Conversations on the Plurality of Worlds). Published in 1686, when the author was only 29, the book became an instant bestseller and is still on sale today. Three different English translations appeared within two years (five more eventually followed, the most recent in 1990). A charming translation is by English poet John Glanvill, *A Plurality of Worlds*, published in 1688 (the latest reprinting was in 1929, a beautifully crafted limited edition by the Nonesuch Press in London; the quotes on these pages are from this edition).

The book was more fiction than physics, but the ideas of other worlds that it displayed were daring, disputable, even disapproved of by the Catholic Church. It was duly placed on the Index of prohibited books a year after publication, removed in 1825 but reinstated in 1900. The Church's disapproval could not dampen the soaring popularity of the book, especially among women. (I have placed a woman in these Conversations, Fontenelle writes in the preface, to encourage women through the example of a woman, 'who without any supernatural parts, or tincture of Learning', never fails to understand what is said to her.)

Fontenelle continued to revise this beloved book, producing 32 editions, the last in 1742, keeping it up to date with the latest scientific discoveries (adjusting the size of Venus or the distance of Saturn, for example, with the most recent astronomical observations).

Written in a playful, whimsical style, *Entretiens* is a delightful, flirtatious conversation between a learned philosopher and a sharp-witted countess that takes place on five consecutive moonlit evenings. The philosopher is convinced that the universe is teeming with intelligent life everywhere. The countess initially scoffs at this bizarre idea, but the philosopher gradually wins her over.

In the first conversation, *Entretiens* introduces the Copernican system: 'Copernicus confounding everything, tearing in pieces the beloved Circles of Antiquity, and shattering their Crystal Heavens like so many Glass Windows: seiz'd with the noble Rage of Astronomy, he snatcheth up the Earth from the Centre of the Universe, sends her packing, and placeth the Sun in the Centre to which it more justly belong …

'Fairly and softly, *saith the Countess*, I fancy you your self are seiz'd with the Noble Fury of Astronomy; a little less Rupture, and I shall understand you the better.'

The second and third conversations are about the Moon, the planets and their inhabitants: 'Fear it not Madam, *said I*, do you think we are the only Fools of the Universe? Is it not consistent with Ignorance to spread it self every where? … Why should Nature be so partial, as to except only the Earth? But let who will say the contrary I must believe the Planets are peopled as well as the Earth.'

The fourth conversation introduces the complex motion of planets in terms of Descartes' vortices: 'Must my Head, *says she, smiling*, turn round to comprehend 'em, or must I become a perfect Fool to understand the misteries of Philosophy.'

In the last conversation, the philosopher takes the final leap and discusses the fixed stars as suns, around which inhabited planets revolve: 'I perceive, *says the Countess*, where you would carry me; you are going to tell me if the fix'd Stars are so many Suns, and our Sun the centre of a Vortex that turns round him, why may not every fix'd Star be the centre of a Vortex that turns round the fix'd Star? Our Sun enlightens the Planets; why may not every fix'd Star have Planets to which they give light?

'You have said it, *I reply'd*, and I will not contradict you.

'You have made the Universe so large, *says she*, that I know not where I am, or what will become of me; what is it all to be divided into heaps confusedly, one among another? Is every Star the centre of a Vortex, as big as ours? Is that vast space which comparehends our Sun and Planets, but an inconsiderable part of the Universe? and are there as many such spaces, as there are fix'd Stars? I protest it is dreadful.

'Dreadful, Madam, *said I*; I think it very pleasant, when the Heavens were a little blue Arch, stuck with Stars, methought the Universe was too strait and close, I was almost stifled for want of Air; but now it is enlarg'd in heigth and breadth, and a thousand & a thousand Vortex's taken in; I begin to breath with more freedom, and think the Universe to be incomparably more magnificent that it was before.

'Well, *says the Countess*, I have now in my Head, the System of the Universe: How learned am I become?'

Entretiens is more than an immortal classic; it is the first popular-science book ever published; a book that not only became popular but also succeeded in making its scientific message overwhelmingly popular.

'The frontier where knowledge ends and ignorance begins'

By the beginning of the 19th century the idea of life on other worlds had entered scientists' consciousness. 'Men of science readily accepted a pleasing hypothesis which was not positively contradicted by evidence from their special studies,' writes Isaac Todhunter in his 1876 book, *William Whewell*. He cites 'almost a solitary expression of a contrary opinion', a flippant remark in *Table Talk*, the conversations of poet Samuel Taylor Coleridge published after his death in 1834: 'I never could feel any force in the arguments for a plurality of worlds, in the common acceptance of that term. A lady once asked me – "What then could be the intention in creating so many great bodies, so apparently useless to us?" I said – I did not know, except to make dirt cheap.'

The name of English polymath William Whewell looms large in 19th-century science (he coined the word 'scientist' to replace the terms then in use: 'natural philosopher' or 'man of science'). In 1853 he published a book, *Of the Plurality of Worlds: An Essay*, in which he contradicted the 'pleasing hypothesis' that extraterrestrial life was rife in the universe. He presented scientific evidence to show the uniqueness of our planet and the improbability of intelligent life elsewhere. 'The discussions in which we are engaged belong to the very boundary regions of science, to the frontier where knowledge ends and ignorance begins,' he wrote.

David Brewster, the Scottish physicist, chiefly remembered today for the invention of the kaleidoscope, accused him of depopulating the heavens. By 1859 five editions (each larger than the earlier) had been published, and their reviews filled the periodicals. The debate that Whewell's book started lasted for more than a decade.

In fact, that debate has never ended. It will not end until the learned countess has contacted the philosopher's other worlds.

CHAPTER 15

FIVE FICTIONAL FACES OF ET

Kepler's dream

Whatever is born from the soil or walks on the soil is of prodigious size. Growth is very quick; everything is short-lived, although it grows to such enormous bodily bulk. They have no settled dwellings, no fixed habitation; they wander in hordes over the whole globe in the space of one of their days, some on foot, whereby they far outstrip our camels, some by means of wings, some in boats pursue the fleeing waters … Most creatures can dive … they live deep down under the water.

Who are they? Klingons from planet Kronos, or Wookiees from planet Kashyyyk, or Daleks from planet Skaro ready to 'EX-TER-MIN-ATE!' earthlings? No, they are giant yet harmless Privolvans from the Moon. They are the first 'aliens' to inhabit the first 'other world' created in a science fiction story, a story dreamed more than 400 years ago by a university student.

Kepler was a 22-year-old student at the University of Tübingen in Germany when, in 1593, he dreamed of writing a book: not an astronomical treatise but a science fiction. The idea occurred to him when he decided to devote his graduate thesis to the question: How would the heavenly phenomena appear to an observer on the Moon?

This question, observes Kepler scholar Max Caspar, was inspired by his enthusiasm for the new astronomy espoused by Copernicus, the man whose prophet he was to become. Caspar notes that the idea of the book 'contains the first germ of a work which we shall come to know as the last of the books he published'.

Twelve years passed before Kepler wrote the first draft, and another 12 before he found time to look at it again. During the last decade of his life, from 1621 to 1630, he began to add copious notes to the slender volume (about 20 pages of a modern book). These notes – 223 in all – are four times as long as the text. The book, written in Latin, was published in 1634, four years after Kepler's sudden death.

Somnium sive Astronomia Lunaris (Dream or Astronomy of the Moon) describes a young boy's trip from Earth (named Volva in the book) to the Moon (Lavinia). The boy, Duracotus, lives with his mother Fiolxhilda on an island. Fiolxhilda is a witch who summons up a demon to propel him to the Moon. Kepler is somewhat vague about the method of propulsion, but he is specific about the distance of the Moon – '50,000 German miles' – which is the correct distance if we use the value of a German mile provided in one of his footnotes.

Unlike the ancient Greeks, Kepler knew that the Earth's atmosphere does not extend as far as the Moon. To enable Duracotus to pass through the thinning atmosphere, he is put to sleep with the aid of opiates and his nostrils stopped with moistened sponges. So that the boy's body is not torn apart by the great acceleration needed to escape Earth, he sits with his arms and legs curled inwards. As he is pulled by the demon towards the Moon, he reaches a point 'where the Moon's magnetic force balances that of the Earth'. When the demon releases him, he falls to the Moon unaided. Here Kepler is groping for gravity, well before Newton saw the proverbial apple fall from a tree and hit upon the idea of it.

'Like spiders they will stretch out and contract and propel themselves forward by their own force – for, as the magnetic forces of the Earth and the Moon both attract the body and hold it suspended, the effect is as if neither of them was attracting it – so that in the end its mass will by itself turn towards the Moon.' That's how he describes the effect of space travel on his hero's body.

'He not only takes it for granted, but, with truly astonishing insights, postulates the existence of "zones of zero gravity" – that nightmare of science fiction,' marvels Arthur Koestler in his magnificent biography of Kepler in *Sleepwalkers* (1959).

When his hero's journey is completed, Kepler proceeds to describe in detail the astronomy, landscape, climate and plant and animal life of both hemispheres of the Moon, called Sublova and Prilova. The story ends when Duracotus is awakened from his dream by a cloudburst.

Somnium skilfully combines a fictional narrative with a treatise on lunar astronomy. It's 'hard' science fiction, the first of its kind. It must be appreciated, as American historian Lewis Mumford says, 'for the audacity of the concept' as well as for its intrinsic merit as a pioneering work of science fiction.

Voltaire's satire

Kepler went to the Moon to see aliens, but Voltaire brought them to Earth in *Micromégas* ('littlebig' in Greek), a whimsical tale of about 6,000 words. His eponymous hero Micromégas, a 36,000-metre tall, handsome giant with a 15,000-metre waist and a 1,700-metre nose is the first science fiction alien to visit our puny planet.

Micromégas has been banished from the star Sirius for 800 years for making 'suspicious, offensive, rash and heretical statements' in a book about little insects less than 30 metres in diameter. He decides to travel 'from planet to planet, with a view to improving his mind and soul'. He has such 'a marvellous acquaintance with the laws of gravitation, and all of the forces of attraction and repulsion' that he travels through the Milky Way 'sometimes by a means of sunbeam, and sometimes with the help of a comet'. He arrives at the planet Saturn and befriends a native, who is only 1,500 metres tall, a mere dwarf beside Micromégas. Together they travel throughout the solar system, and when they reach Earth the Saturnian complains that 'this globe is so ill-constructed, so irregular, and so ridiculously shaped … I cannot suppose any sensible people should wish to occupy such a dwelling'.

Like *Candide*, Voltaire's most popular work, *Micromégas* is a satire. Published in 1752, it makes satirical observations on the pretensions of humans. The visitors to Earth marvel at its smallness and its 'microscopic' inhabitants. This was a revolutionary thought in Voltaire's time, when Earth was believed to be the centre of the universe and humans its perfect creatures.

The Sirian, who has a thousand senses as opposed to the poor Saturnian's 72, notices a shipload of philosophers who are returning from a voyage to the polar circle. Micromégas picks up the ship and places it in the hollow of his hand. He then examines 'the mere mites' with a microscope he has brought with him. The two travellers are amazed to find that the 'invisible insects' can speak and even have the capacity to reason. Once they get over their amazement, they start lengthy conversations with the philosophers.

'O intelligent atoms … you must doubtless taste joys of perfect purity on your globe; for, being encumbered with so little matter and seeming to be all spirit, you must pass your lives in love and meditation – the true life of spiritual beings.' The philosophers all shake their heads and one, more frank than the others, replies: 'We have more matter than we need, the cause of much evil if evil proceeds from matter; and we have too much mind, if evil proceeds from mind. For instance, at this very moment there are 100,000 fools of our species who wear hats, slaying 100,000 fellow creatures who wear turbans, or being massacred by them.'

Micromégas shudders and asks the philosopher the cause of such horrible quarrels. 'The dispute concerns a lump of clay,' replies the philosopher, 'no bigger than your heel.'

We run the risk of being called bore if we tell you more, as Voltaire (pseudonym for François Marie Arouet) once said: 'The secret of being a bore is to tell everything.' ('*Le secret d'ennuyer est celui de tout dire.*')

Hoyle's cloud

In Fred Hoyle's science fiction novel, *The Black Cloud*, the alien is one giant leap of imagination. It takes on the form of a cloud of interstellar gas – as large as the distance from the Sun to Earth and as massive as Jupiter – that can move as it wishes from star to star. This 500-million-year-old intelligent organism has a 'brain' that consists of neurological structures built from complicated molecules. The creature can increase the capacity of its brain by extending these structures and by learning to use them in the best way to solve problems as they arise.

Scientists notice the cloud when it approaches the solar system on a course that is predicted to bring it between the Sun and Earth, causing a global catastrophe. They discover that the cloud is alive and try to communicate with it. Once they establish contact, they warn the cloud that certain aggressive governments are trying to destroy it by hydrogen bomb rockets, an act 'sillier than trying to kill a rhino with a toothpick'. The cloud reverses the course of the rockets so that they 'hit the exact points they started from', destroying many cities. Eventually, the cloud decides to leave, but not before it has enjoyed Micromégas-like conversations with scientists and has expressed its views on matters ranging from the origin of headaches to the nature of intelligence. Some samples:

On why intelligent animals are unusual on planets: 'Living on the surface of a solid body, you are exposed to a strong gravitational force. This greatly limits the size to which your animals can grow and hence limits the scope of your neurological activity. It forces you to possess muscular structures to promote movement ... Indeed, your very largest animals have been mostly bone and muscles with very little brain ... By and large, one only expects intelligent life to exist on a diffuse gaseous medium, not on planets at all.' On reproduction: 'I can live indefinitely, you see. Therefore, I am not under the necessity, as you are, of generating some new individuals to take over after my death.'

In the book's preface, Hoyle says that he hopes that his scientific colleagues will enjoy this 'frolic'. He has also a disclaimer to the effect that none of the characters should be identified with the author. 'Be that as it may, the central character, Chris Kingsley, is a professor of astronomy at Cambridge, and Hoyle's handling is distinctly autobiographical,' says Simon Mitton in *Conflict in the Cosmos: Fred Hoyle's Life in Science* (2005). '*The Black Cloud* reads authentically because Hoyle sets the action in places he knew well: Caltech, Palomar Observatory, the RAS, and Cambridge. His scientists are doing real science, not fantasy science: We see them using calculus, scientific diagrams, and the EDSAC-1 computer – all graphically described.'

The Black Cloud was Hoyle's first science fiction novel and was a sensational bestseller when published in 1957. It has now achieved classic status. Hoyle wrote and co-authored many other science fiction books before his death in 2001, but it is *The Black Cloud* for which he is chiefly remembered as a science fiction writer.

Hoyle, an astrophysicist, was a staunch supporter of the steady-state theory and never gave up his belief in it: the universe has no beginning and will have no end. This theory is now considered flawed, and the big bang theory is widely accepted: the universe began when a single point of infinitely dense and infinitely hot matter exploded spontaneously. The name 'big bang' was given by Hoyle. It was meant to be a put-down when he first used it scornfully in a radio talk in 1950.

Lem's brooding ocean

The alien – an ocean that lives, thinks and acts – in Stanislaw Lem's science fiction novel *Solaris* is another giant leap of imagination. But unlike the black cloud, this ocean is not 'talkative'. Humans cannot communicate with it; they cannot even comprehend its intelligence.

The distant planet Solaris orbits two suns – a red sun and a blue sun. Its sole organic inhabitant is an ocean that covers the entire planet. The ocean expresses itself by electrical and magnetic impulses in a more or less mathematical language. These 'monologues' unfolding in the depths of the ocean's brain are incomprehensible to scientists who have been studying it from a space station orbiting the planet. They are convinced that they are confronted with 'a monstrous entity endowed with reason, a protoplasmic ocean-brain enveloping the entire planet and idling its time away in extravagant theoretical cogitation about the nature of the universe'.

The ocean regularly produces gigantic constructions in various shapes from a substance resembling a yeasty colloid. The scientists describe them in artificial, linguistically awkward terms such as 'tree-mountains', 'fungoids', 'symetriads', 'asymetriads' and 'vertebrids'. Were the constructions the 'taciturn' ocean's way of expressing itself?

'I only wanted to create a vision of a human encounter with something that certainly exists, in a mighty manner perhaps, but cannot be reduced to human concepts, ideas or images,' Lem explained on his website 41 years after the publication of his book. 'The ocean neither built nor created anything translatable into our language that could have been "explained in translation". Hence a description had to be replaced by analysis – obviously an impossible task – of the internal workings of the ocean's ego. This gave rise to *symetriads*, *asymetriads* and *mimoids* – strange semi-constructions scientists were unable to understand; they could only describe them in a mathematically meticulous manner.'

Kris Kelvin, the protagonist of *Solaris*, is a psychologist who arrives at the space station to determine whether research into Solaris should be terminated for want of progress. But he finds the space station all but deserted, it straggling crew seemingly haunted by hallucinations of 'phantoms' from their respective pasts. In the middle of the night, Kelvin himself is visited by an exact replica of his wife, Rheya, who had committed suicide after being estranged from him.

Was the ocean using these phantoms created from neutrinos to pry into scientists' minds? Yes, says Lem, but it did it in a radical way. 'It penetrated the superficial established manners, conventions and methods of linguistic communication, and entered, in its own way, into the minds of the people of the Solaris station and revealed what was deeply hidden in each of them: a reprehensible guilt, a tragic event from the past suppressed by the memory, a secret and shameful desire.'

Beyond the issue of communicating with an alien life form, *Solaris* is also about whether contact should be made. 'We don't want to conquer the cosmos, we simply want to extend the boundaries of Earth to the frontiers of the cosmos,' sermonises one of the scientists based on the Solaris station. 'We think of ourselves as the Knights of the Holy Contact. This is another lie. We are only seeking Man. We have no need of other worlds. We need mirrors. We don't know what to do with other worlds.'

Lem, who died at 84 in 2006, is one of the greatest science fiction writers of all time. He wrote dozens of books, all in Polish, but *Solaris* is his masterpiece. Described as 'intellectual science fiction', it was first published in 1961 and has been translated into more than 40 languages. In 1972 the legendary Russian director Andrei Tarkovsky adapted *Solaris* into a film of the same title, which was described at the Cannes Film Festival as 'the most intelligent and insightful film in the history of science fiction movies'. Hollywood director Steven Soderbergh's *Solaris* (2002) is a well-crafted remake of the Russian classic.

Sagan's contact

In *Contact*, Carl Sagan creates a 'Knight of the Holy Contact'. Dr Ellie Arroway, Sagan's protagonist, receives a coded message in a radio signal: the Message is a plan for building a machine to transport five passengers to its benevolent sender on the star Vega, 26 light-years away.

Arroway, 'a pretty woman with evident scientific competence who forcefully expresses her views', is not only faced with the scientific and technical problems of building the Machine; she has to deal with political ('who speaks for Earth?'), religious ('is the purpose of the Message divine or satanic?'), economic ('fears that building the Machine would ruin the world economy') and gender ('some would interrupt her or pretend not to hear her') issues.

There is 'narrow vocational disquiet' as well: 'There were other intelligent beings in the universe. We could communicate with them. They were probably older than we, possibly wiser … So, the specialists in every subject began to worry. Mathematicians worried about what elementary discoveries they might have missed. Religious leaders worried that Vegan values, however alien, would find ready adherents, especially among the uninstructed young …'

And there are cultural subtleties to be taken care of. At the World Message Consortium, the Chinese delegate stresses: 'We have received an invitation. A very unusual invitation. Maybe it is to go to a banquet. The Earth has never been invited to a banquet before. It would be impolite to refuse.' Who would refuse an invitation to a sumptuous Vegan banquet?

Arroway accepts the invitation and the challenge of confronting religious opposition, political rivalry and scientific jealousy. The Machine is eventually built and makes a trip to Vega, but we never find out what the Vegans look like. But you hardly expect to see humanoid-like aliens in his only novel, because Sagan believed that our own forms were such an accident of numerous random evolutionary events that the odds of aliens looking humanoid were small.

Like Sagan's immensely successful *Cosmos* – the 1980 book and TV series – *Contact* makes science accessible. It's filled with scientific facts and fervour, which makes it utterly believable. *Contact* was published in 1985, and the film of the same name, directed by Robert Zemeckis, was released in 1997, a year after Sagan's death.

'We inhabit an insignificant planet of a hum-drum star lost in a galaxy tucked away in some forgotten corner of a universe in which there are far more galaxies than people,' Sagan said in *Cosmos*. 'We make our world significant by the courage of our questions, and by the depth of our answers.'

The story of the search for extraterrestrials continues. To paraphrase Sagan, it will end only when humanity's courageous questions have the right depth of answers. We may soon learn about the discovery of microbial life on another world, but will the question of intelligent life on other worlds ever have the right depth of answer?

ABOUT THE AUTHOR

Surendra Verma is a science writer based since 1970 in Melbourne, Australia. He has published numerous popular science books in Australia, the UK, the US and India, which have been translated into 14 languages other than English. His recent books include:

The Mystery of the Tunguska Fireball

Why Aren't They Here? The Question of Life on Other Worlds

The Cause of Mosquitoes' Sorrow: Beginnings, Blunders and Breakthroughs in Science

The Little Book of Scientific Principles, Theories & Things

The Little Book of Maths Theorems, Theories & Things

The Little Book of Unscientific Propositions, Theories & Things

The Little Book of the Mind: How We Think and Why We Think

Learn & Unlearn: The novel way to rethink the things that matter in your life

The Little Big Book of Science in 100 Words

+ a children's book

Who Killed T. Rex? Uncover the mystery of the vanished dinosaurs